Particle Sizing and Characterization

ACS SYMPOSIUM SERIES **881**

Particle Sizing and Characterization

Theodore Provder, Editor
Eastern Michigan University

John Texter, Editor
Eastern Michigan University

**Sponsored by the
ACS Division of Colloid and Surface Chemistry**

American Chemical Society, Washington, DC

Library of Congress Cataloging-in-Publication Data

Particle sizing and characterization / Theodore Provder, editor, John Texter, editor ; sponsored by the ACS Division of Colloid and Surface Chemistry.

    p. cm.—(ACS symposium series ; 881)

    Includes bibliographical references and index.

ISBN-13 978-0-84-123859-6
ISBN 0–8412–3859–6

    1. Particle size determination—Congresses

    I. Provder, Theodore, 1939- II. Texter, J. (John) III. American Chemical Society. Division of Colloid and Surface Chemistry. IV. American Chemical Society. Meeting (224th : 2002 : Boston, Mass.) V. Series.

TA418.8.P347   2004
620′.43—dc22   2003063662

The paper used in this publication meets the minimum requirements of American National Standard for Information Sciences—Permanence of Paper for Printed Library Materials, ANSI Z39.48–1984.

PRINTED IN THE UNITED STATES OF AMERICA

Foreword

The ACS Symposium Series was first published in 1974 to provide a mechanism for publishing symposia quickly in book form. The purpose of the series is to publish timely, comprehensive books developed from ACS sponsored symposia based on current scientific research. Occasionally, books are developed from symposia sponsored by other organizations when the topic is of keen interest to the chemistry audience.

Before agreeing to publish a book, the proposed table of contents is reviewed for appropriate and comprehensive coverage and for interest to the audience. Some papers may be excluded to better focus the book; others may be added to provide comprehensiveness. When appropriate, overview or introductory chapters are added. Drafts of chapters are peer-reviewed prior to final acceptance or rejection, and manuscripts are prepared in camera-ready format.

As a rule, only original research papers and original review papers are included in the volumes. Verbatim reproductions of previously published papers are not accepted.

ACS Books Department

Contents

Fractionation Methods and Applications: Ultracentrifugation, Capillary Hydrodynamics Fractionation, and Field-Flow Fractionation

Electrophoretic, Electrokinetic, and Acoustic Attenuation Methods

Indexes

Preface

The previous three books in this series (*1–3*) emphasized the continuing revitalization in particle size measurement and particle characterization due to advances in electronic and computer technology as well as in market forces generated by customer needs for improved measurements of particle size and characterization of particles. Many commercial product properties are dependent on the particle size and particle properties of the components in the product formulation whether the product final form is a solid, dispersion, or fine emulsion. To gain a competitive advantage, manufacturers of such products need to understand the relationship of particle size and properties to product end-use properties.

In addition the growth of nanoparticle materials technology significantly is impacting product properties in such diverse product areas as coatings, inks, cosmetics, printing media, imaging, paper products, diagnostic testing as well as devices and displays. Nanoparticle materials characterization needs also are significant driving forces for improved particle size measurement and characterization.

The symposium on which this book is based reflects current activities in particle size measurement, characterization techniques, and applications brought about by the driving forces mentioned above. The topics in this book are divided into three sections. The first section covers methods based on various forms of light scattering, including applications involving dynamic light scattering, multiple light scattering, turbidity, and fluorescence (confocal microscopy) methods. The second section covers fractionation methods including ultracentrifugation and column chromatography techniques that are focused on nanoparticles. The third section covers the field of particle characterization involving electrophoretic, electrokinetic, and acoustic attenuation methods to characterize product surfaces and interfaces, particle–particle interactions and concentrated dispersions.

It is expected that improvements in instrumentation capability and particle measurement and characterization will continue to be manifested in commercially available instrumentation. The editors expect that the previous three books in this series as well as this volume will be useful guides to those working in the field of particle size measurement and particle characterization and will motivate further activity among experienced practitioners in the field. We thank the authors for their effective oral and written communications and the reviewers for their helpful critiques and constructive comments.

References

1. *Particle Size Distribution: Assessment and Characterization*; Provder, T., Ed.; ACS Symposium Series 332; American Chemical Society: Washington, DC, 1987.
2. *Particle Size Distribution II: Assessment and Characterization;* Provder, T., Ed.; ACS Symposium Series 472; American Chemical Society: Washington, DC, 1991.
3. *Particle Size Distribution III: Assessment and Characterization;* Provder, T., Ed.; ACS Symposium Series 693; American Chemical Society: Washington, DC, 1998.

Theodore Provder, Director
Coatings Research Institute
Eastern Michigan University
430 West Forest Avenue
Ypsilanti, MI 48197

John Texter, Professor
Department of Interdisciplinary Technology
Eastern Michigan University
Ypsilanti, MI 48197

Light Scattering Methods and Application

Chapter 1

PCS Particle Sizing in Turbid Suspensions: Scope and Limitations

Frank Scheffold[1,*], Andrey Shalkevich[1], Ronny Vavrin[1], Jérome Crassous[2], and Peter Schurtenberger[1]

[1]Physics Department, Soft Condensed Matter Group, University of Fribourg, CH–1700 Fribourg, Switzerland
[2]Laboratoire de Physique, 46 allée d'Italie, Ecole Normale Superieure de Lyon, 69364 Lyon cedex 7, France
*Corresponding author: Frank.Scheffold@unifr.ch

Recent advances in PCS particle size characterization in turbid suspensions are discussed. We show two complementary routes to overcome the limitations of traditional PCS with respect to multiple scattering of light. The first one, called 3D dynamic light scattering (3DDLS) aims to suppress multiple scattering with a two detector cross correlation scheme. Combined with a parallel processing of experiments at different angles such a setup may significantly improve resolution and stability for the analysis of complex particle size distributions in turbid suspensions. The second technique, diffusing wave spectroscopy (DWS), is shown to be a versatile method to obtain information about the average particle size based on the analysis of diffusely reflected light.

Photon correlation spectroscopy has become an important tool for particle sizing in suspensions and slurries over the last 25 years [1-4]. The technique exploits the fact that light scattered from a (dilute) suspension of particles fluctuates with a characteristic time scale inversely proportional to the particle diffusion constant. It can provide information about particles with sizes ranging from a few nanometers to several micrometers. However the range of application of standard PCS is restricted to single scattering which severely limits its use for industrially relevant systems[5,6]. Particularly for larger particles with high scattering contrast the technique can only be applied to very low particle concentrations, and a large variety of systems are therefore excluded from investigations with PCS. In this paper we discuss two alternative routes to overcome these limitations and to perform PCS in turbid suspensions. The first technique, called 3D-Dynamic Light Scattering (3DDLS), allows the suppression of multiple scattering [7-12] while the second technique, Diffusing Wave Spectroscopy (DWS), follows the fluctuations of diffuse light [4,13-17]. Both techniques analyse the temporal fluctuations of the scattered light, as with PCS, however they rely on sophisticated schemes to overcome the undesired multiple scattering effects in traditional PCS.

Photon Correlation Spectroscopy

In a PCS experiment laser light is incident on a scattering cell of dimension L containing a scattering medium. In the case of conventional PCS, also called dynamic light scattering (DLS), the scattering mean free path $l \gg L$ and therefore the transmission in line of sight has to be high, $\exp(-L/l) > 95\%$[1,2,18]. However, for many suspensions of industrial relevance and with larger colloidal particles this is often not the case. Thus, here we have to find alternatives to either omit multiple scattering or then to suppress. Before we show how this can be done in a cross correlation experiment, we briefly review the basis of PCS.

Dilute or Weakly Scattering Suspensions

In the most simple case we can consider a system of N independent point scatterers illuminated with laser light. The scattered light is then detected at a certain angle θ, hence incoming and scattered wavevector (with $|\mathbf{k}_i| = |\mathbf{k}_s|$, elastic scattering) define the scattering vector $\mathbf{q} = \mathbf{k}_i - \mathbf{k}_s$, $|\mathbf{q}| = 2|\mathbf{k}| \sin(\theta/2)$. The intensity fluctuations due to the motion of the colloidal particles are commonly analysed by generating an intensity autocorrelation function (IACF) [19]

$$G_2(q,t) = \langle I(q,0)I(q,t)\rangle_T \qquad (1)$$

where the average is a time average. For ergodic (fluid-like) systems the IACF is directly related to the normalized field autocorrelation function $g_1(t)$ through the Siegert Relation

$$g_2(q,t) = \frac{\langle I(q,0)I(q,t)\rangle_T}{\langle I(q,0)\rangle_T^2} = 1 + \beta|g_1(q,t)|^2 \tag{2}$$

where the normalized field autocorrelation function $g_1(t)$ is given by

$$g_1(t) = \frac{\langle E(q,0)E(q,t)\rangle_T}{\langle E(q,0)^2\rangle_T} \tag{3}$$

The intercept β primarily dependends on the detector optics, and the ideal value can (almost) be reached with single mode fibers for the detection optics[19]. For non-interacting and monodisperse particle suspensions in the single scattering limit $g_1(t)$ decays like a single exponential

$$g_1(q,t) = e^{-D_0 q^2 t} \tag{4}$$

which combined with the Stokes-Einstein relation $D_0 = k_B T/6\pi\eta R_h$), where η is the solvent viscosity and R_h is the hydrodynamic radius, represents the basis for the application of PCS in particle sizing. For polydisperse non-interacting spherical particles with a number-weighted size distribution $N(r)$, eqn. 4 has to be written as

$$g_1(q,t) = \frac{\int_0^\infty N(r)A^2(q,r)e^{-D_0(r)q^2 t}dr}{\int_0^\infty N(r)A^2(q,r)dr} \tag{5}$$

where $A(q,r)$ is the scattering amplitude and $D_0(r)$ the diffusion coefficient of a particle with radius r. Eqn. (5) is often written as

$$g_1(q,t) = \int_0^\infty G(\Gamma)e^{-\Gamma t}d\Gamma \tag{6}$$

where $\Gamma = D_0 q^2$ and $G(\Gamma)$ is the intensity-weighted size distribution, which for larger particles with anisotropic scattering depends both on the particle size distribution $N(r)$ as well as on the scattering angle θ.

Based on eqn. (6) one can then try to obtain the full size distribution N(r) from PCS experiments with polydisperse colloidal suspensions. However, the basic problem in the data analysis results from the ill-conditioned nature of the inversion of eqn (6), i.e. the fact that even a small amount of noise on the experimental correlation function, together with the limited accessible τ-range, may greatly distort G(Γ). Thus, there exists an entire set of solutions to eqn (6) that lie wihtin the experimental noise level ε. While some of them can be discarded immediately because they have no 'physical significance' (such as those with negative amplitudes), we still end up with the substantial problem of having to choose the 'correct' or most probable solution among all the feasible ones. There exist numerous approaches to this problem such as the often used program CONTIN. However, it is clear that there will always be limits to the use of PCS, and in particular in many suspensions of practical importance we are also confronted with the problem of multiple scattering.

3D Dynamic light scattering (3DDLS)

The application of PCS to many systems of scientific and industrial relevance has often be considered as too complicated due to the very strong multiple scattering frequently encountered. The interpretation of a SLS/DLS experiment becomes exceedingly difficult for systems with non-negligible contributions from multiple scattering. Particularly for larger particles with high scattering contrast, this limits the technique to very low particle concentrations, and a large variety of systems are therefore excluded from investigations with dynamic light scattering. However, Schätzel has demonstrated that it is possible to suppress contributions from multiple scattering from the measured photon correlation data [7]. During the last few years a number of different theoretical and experimental approaches to this problem have appeared [6]. The general idea is to isolate singly scattered light and suppress undesired contributions from multiple scattering by using a so-called cross correlation scheme. This can be achieved by performing two scattering experiments simultaneously on the same sample (with two mutually incoherent laser beams, initial wave vectors k_{i1} and k_{i2}, and two detectors positioned at final wave vectors k_{f1} and k_{f2}) and cross-correlating the signals seen by the two detectors. Provided the two light scattering experiments share the same scattering volume and have identical scattering vectors $q_1 = q_2$ (or $q_1 = -q_2$) but use different geometries, only singly scattered light will produce correlated intensity fluctuations on both detectors. In contrast, multiply scattered light will result in uncorrelated fluctuations that contribute to the background only due to the fact that it has been scattered in a succession of different q vectors. The corresponding relationships between the field autocorrelation function $g_1(t)$ and the measured auto- ($G_{11}^{(1)}(t)$) and cross-correlation ($G_{12}^{(1)}(t)$) functions, where $^{(1)}$ indicates singly scattered photons, can then be written as

$$G_{11}^{(1)}(t) = \left\langle \left(I_1^{(1)}\right)^2 \right\rangle \left(1 + \beta_{11} |g_1(q,t)|^2\right) \tag{7}$$

and

$$G_{12}^{(1)}(t) = I_1^{(1)} I_2^{(1)} \left(1 + \beta_{12} |g_1(q,t)|^2\right) \tag{8}$$

where I_1 and I_2 are the average scattered intensities seen by detectors 1 and 2, respectively. Whereas the intercept β_{11} depends primarily as stated above on the detection optics and has an ideal value of 1, the cross-correlation function β_{12} is in addition reduced by phase mismatch and optical misalignement (i.e., match of the two scattering volumes). For single scattering, both auto- and cross-correlation therefore yield the same information. However, in the case of multiple scattering, the situation changes. In the auto-correlation experiment the multiply scattered photons contribute to $G_{11}^{(1)}(t)$ as well and make interpretation of the data and a deduction of $g_1(t)$ very difficult if not impossible. However, in the cross-correlation experiment only singly scattered light will produce correlated intensity fluctuations on both detectors. In contrast, multiply scattered light will result in uncorrelated fluctuations that contribute to the background only due to the fact that it has been scattered in a succession of different $\mathbf{q}$ vectors, and the contributions from multiple scattering to the signal are suppressed. This leads to the following expression for $G_{12}^{(1)}(t)$

$$G_{12}^{(1)}(t) \approx \left\langle I_1 \right\rangle \left\langle I_2 \right\rangle + \beta_{12} \left\langle I_1^{(1)} \right\rangle \left\langle I_2^{(1)} \right\rangle |g_1(q,t)|^2 \tag{9}$$

where $\langle I_j \rangle$ is the average intensity at detector j and $\langle I_j^{(1)} \rangle$ contains the singly scattered light only. Such a cross-correlation experiment will thus provide us with $g_1(t)$ even for turbid suspensions, and multiple scattering will be visible in the decrease of the intercept.

A particularly interesting cross-correlation experiment is the so-called 3d coding described in in [7,9]. Two laser beams $\mathbf{k}_{i1}$ and $\mathbf{k}_{i2}$ are crossing each other at the scattering volume (see figure 1). The two incident and the two detected light paths are placed at an angle $\delta/2$ above and below the plane of symmetry of the experiment to obtain identical scattering vectors $\mathbf{q}_1 = \mathbf{q}_2$. In the 3D experiment the two scattering processes 1-1 and 2-2 thus have the same q, whereas the two other scattering processes 1-2 and 2-1 detected in this experiment have different scattering vectors. These additional scattering processes will therefore contribute to the background only, which means that the maximum intercept for 3DDLS is only one quarter of the value obtained for auto-correlation or other cross-correlation schemes such as the two-colour method, i.e. $\beta_{12,ideal} = 0.25$.

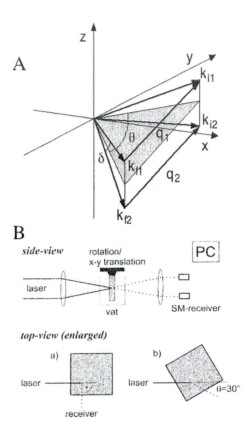

Figure 1. (A): wave-vector arrangement in the 3d cross-correlation set-up; (B): Schematic view of a 3DDLS setup. Two parallel beams are focused onto the scattering cell by a lens. An identical lens is used for the detection side, and the scattered light is collected using single mode fibers. The top view shows how rectangular scattering cells can be used combined with a θ - 2θ geometry (b) in order to work with a minimized optical path length

A schematic layout of a 3DDLS experiment is shown in figure 1. A laser beam is split into two parallel beams, which are then focused onto the scattering cell by a lens. An identical lens is used for the detection side, and the scattered light with wave vectors k_{f1} and k_{f2} is collected using single mode fibers. For turbid colloidal suspensions, due to the strong attenuation of the singly scattered intensity which essentially decays like $I^{(1)} \sim \exp(-s/l)$, where s is the optical path across the cell for singly scattered light and l is the scattering mean free path, which may easily be less than 1 mm for dense colloidal suspensions, the size of the scattering cell becomes important. In order to avoid a strong reduction of the singly scattered light and the subsequent reduction in signal, the light path should be of the order of l. In principle this could be achieved best by using flat cells. However, since these cells will then be used in general with beams at non-normal incidence angles, this will lead to considerable beam deflection and displacement. This problem can be overccome by implementing a different approach outlined in figure 1. One can use square cells with 10 mm path length that are of superior optical quality for experiments at low scattering angles, and position them such that the scattering volume is located in a corner of the cell to have short optical path lengths. For experiments at scattering angles different from 90° we can then turn the cell in a so-called θ-2θ geometry, where the sample is rotated by half the rotation angle (90° - θ) in order to recover a symmetrical situation in which the displacement of the incident and scattered beam almost cancel l [9].

Typical results from a 3DDLS measurement with turbid suspensions can be seen in figure 2. Shown are data obtained with a suspension of PMMA particles in decalin with a mean radius of R_h = 136 nm. Measurements were made at increasing concentrations of PMMA from approx. 0.5% to 8.5% by weight. Also shown are photographs of the samples that demonstrate the increasing turbidity. The influence of multiple scattering can be seen directly from the decreasing intercept in fig. 2a, but when rescaled with their intercept the correlation functions all collapse onto a single curve (fig. 2b), thus demonstrating that the 3D instrument provides the correct $g_1(q,t)$ even for highly turbid samples.However, while it is clear that 3DDLS enormously expands the range of samples that can be analyzed with PCS, the possibility to correctly extract the intensity ($G(\Gamma)$) or number weighted ($N(r)$) size distribution through an inversion of eqn. 6 remains questionable due to the inherently reduced signal-to-noise ratio for highly turbid samples. We have thus also tried to optimize the Laplace inversion routine by fully taking advantage of the fact that the angular dependence of $G(\Gamma)$ carries additional information that can be used.

Multi-angle Laplace inversion

In multi-angle scattering experiments different autocorrelation functions are measured at different scattering angles on the same experimental system. It has previously been pointed out that such a technique increases significantly the available resolution on particle size volume distributions [20-24]. The experimental data are analyzed with the constraint that they are performed at known angles. However those methods assume implicitly an accurate knowledge of the static scattering intensities. Here we first investigate the importance of the static measurement. Based on this, a simple method of analyzing multiangle dynamic

10

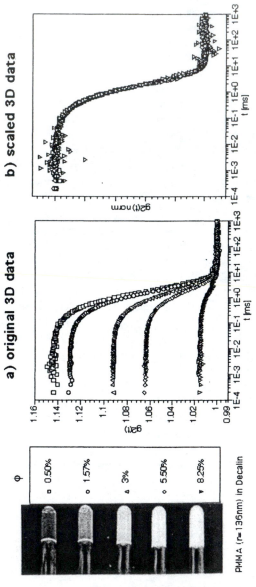

Figure 2. Photographs of PMMA samples measured at different concentrations with 3DDLS together with the thus obtained intensity cross-correlation functions (a). Also shown are the cross correlation functions normalized with their corresponding intercept values (b).

light scattering data is presented. We then test this method experimentally on various mixtures of latex spheres, first highly diluted samples with single scattering only, and then with turbid samples using the 3D instrument with its ability to suppress multiple scattering at the expense of a reduced intercept.

We have already seen that for polydisperse systems the normalized field autocorrelation function $g_1(t)$ is given by eqn. (5). The integrals in equation (5) are, for practical use, written in a discrete form. For this, a discrete preselected set of radii [r_j; $j = 1,...,N$] is chosen, and [c_j; $j = 1,...,N$] are then the set of related (unknown) coefficients:

$$g_1(\theta,t) = \frac{\sum\limits_{j=1}^{N} c_j A^2(\theta,r_j) e^{-\Gamma(\theta,r_j)t}}{\sum\limits_{j=1}^{N} c_j A^2(\theta,r_j)} \tag{10}$$

In a multiangle dynamic light scattering experiment, P different correlations functions are measured at scattering angles $(\theta_1,...,\theta_P)$ and every correlation function is measured at M delay times $(t_1,....,t_M)$. The practical problem of the inversion of experimental data is to find a set [c_j] which fulfills eqn. (10) for every angle θ_j and every delay time t_n.

Since the scattered amplitude depends strongly on the scattering angle, the weight of the contribution from a particle with a given radius also depends strongly on the scattering angle. For example, a suitable selection of scattering angles may be done in such a way that a given particle size dominates in equation (10). Therefore the variation of the scattered intensities are implicitly present in this system of equations. It should be stressed that the normalization of the correlation function in eqn. (3) removes any explicit variations of the static scattered intensities at angles θ_n. In other words, a resolution of this system may be in principle done with or without the extra constraint that the measured static intensities are $I^{(m)}(\theta_n) = \sum_{j=1}^{N} c_j A^2(\theta_n,r_j)$.

The importance of the measured static intensities may be weighted in eqn. (10) with the help of an extra positive parameter λ in such a way that the denominator of eqn. (10) becomes :

$$\sum_{j=1}^{N} c_j A^2(\theta_n,r_j) = (1-\lambda)\sum_{j=1}^{N} c_j A^2(\theta_n,r_j) + \lambda I^{(m)}(\theta_n) \tag{11}$$

with this, eqn. (3) then becomes

$$\lambda g_1(\theta_n,t) = \sum_{j=1}^{N} c_j \frac{A^2(\theta_n,r_j)}{I^{(m)}(\theta_n)}\left((\lambda-1)g_1(\theta_n,t)+e^{-\Gamma(\theta_n,r_j)t}\right) \tag{12}$$

where we can discuss some special cases for eqn. 12:

(i) $\lambda = 0$ leads to eqn. 5, i.e. the measurement of the static intensity is not used in the resolution of the system.

(ii) $\lambda = 1$: In this case the autocorrelation functions are normalized with the measurement of the static scattering experiment.

(iii) $\lambda \to \infty$: eqn. (12) becomes $I^{(m)}(\theta_n) = \sum_{j=1}^{N} c_j A^2(\theta_n, r_j)$. It is the problem of finding a particle distribution [c_j] in agreement with the static measurement only.

Hence λ is a control parameter which allows us to take more or less into account the measurement of the scattered intensity. The practical weight of the static measurement in the inversion of eqn. (10) and consequently the optimal value for λ is expected to be dependent on the details of the experimental set-up. For example, the optimal value should be greater in an experimental system designed for both static and dynamic experiments than for an experimental system designed mainly for dynamic measurements. The procedure to fix the value of λ is reported below in the experimental test of the method.

For practical implementation, eqn. (12) is written in a matricial form

$$\mathbf{y} = \mathbf{Ac} \qquad (13)$$

where $\mathbf{y}$ is a M.P x 1 vector of component $\lambda I^{(m)}(\theta_n) g_1(\theta_n, t_j)$ for $n = 1$ to P and $j = 1$ to M, $\mathbf{A}$ is a M.P x N matrix of component $A^2(\theta_n, r_i)((\lambda - 1)g_1(\theta_n, t_j) + e^{-\Gamma(\theta_n, r_i)t})$ for $n = 1$ to P, $j = 1$ to M, and $i = 1$ to N, and $\mathbf{c}$ is a N x 1 vector of component c_i for $i = 1$ to N. The ordinary least squares solution minimizes the weighted sum of squared residuals

$$\chi^2 = \sum (\mathbf{y}_{j,n} - (\mathbf{Ac})_{j,n})^2 W_{j,n} \qquad (14)$$

where the weight $W_{j,n}$ is related to the uncertainty on the experimental data g_1 and are set, in accordance with [25] to $W_{j,n} = g_1(\theta_n, t_j)^2/(1 + g_1(\theta_n, t_j)^2)$. The numerical solution is performed with a nonnegativity constraint, with the use of the routine NNLS of Lawson and Hanson[26]. The Mie scattering functions for the calculation of the scattering amplitudes $A(\theta_n, r_j)$ are calculated using the routines of Barber and Hill[27].

Multi-angle Laplace inversion applied to dilute suspensions

In order test the performance of the multiangle Laplace inversion routine we first performed light scattering experiments with dilute suspensions without any multiple scattering. We used different mixtures of almost monodisperse latex spheres (Surfactant free latex spheres from IDC Portland). A first set of samples consisted of mixtures of two narrowly distributed latex particle suspensions (r_1 = 60 nm and r_2 = 350 nm) with four different volume ratios $V(r_1)/V(r_2)$ = 10:1, 1:1, 1:10 and 1:100. A second set of samples consisted of four quasicontinuous distributions obtained from mixtures of narrowly distributed latex of radius r = 55; 150; 200; 260; 350; 400 nm with various volume fractions. All solutions were dispersed in water (Milli-Q water purification system from Millipore) up to a typical solid fraction 10^{-5}. Experiments were carried out using an ALV-5000F goniometer system, equipped with an argon-ion laser (Coherent Innova 300, λ_0 = 488 nm). All measurements were carried out at 25.0 ± 0.1 °C. The correlation functions and the static data were measured successively at 7 angles linearly spaced in the

range 30-120°, and the duration of each measurement was 2 minutes. The laser intensity was adjusted such that the count rate at θ = 30° was roughly 200'000 counts per second. Analysis of the experimental data was performed on all systems with a fixed set of logarithmically spaced radii r_j in the range 1-4000 nm. In a preliminary analysis, the optimal values for the number of basis functions N and the parameter λ have been determined. For this, the solution of the system of equations (13) is performed for various values of λ and N. For every set of pairs (λ,N) the particle number distribution [c_i] is obtained, the calculated values of the corresponding correlation functions are reconstructed using (12), and the rms error between calculated and measured correlation

functions $\chi^2 = \sum_{j,n} \left(g_{1;j,n}^{calc} - g_{1;j,n}^{meas} \right)^2 W_{j,n}$ is then obtained. For high values of

N, corresponding to high resolution in the particle number distribution, this error is independent of the value of the parameter λ and is roughly the same as the computed error if correlation functions at different angles are fitted independently. The measured correlation functions are thus correctly evaluated, but in return the resulting particle size distributions exhibit spurious peaks. Such stability problems that arise when the resolution is increased beyond what is allowed by the signal to noise ratio have been reported in both dynamic [28] and static[29] light scattering experiments. For a lower resolution, the choice of λ, i.e. the weight attributed to the static part of data, strongly affects the rms error. For high values of λ, the rms error χ^2 is very important, and decreases regularly with λ. However at too small values of λ, the corresponding rms values increase considerably. This behavior is expected, since setting λ = 0 in eqn. (13) corresponds to finding vectors in the null space of **A**, an operation for which the routine used to perform the inversion of eqn. (13) is not designed. This dependence is systematic for the analysis of every experimental system, with roughly the same numerical values of N and λ. The number N of basis functions is chosen large enough in order to guarantee a sufficient resolution of the particle distributions, and the corresponding value of λ is chosen such that the rms error is minimal. In practice, all the experimental data were analysed with the same numerical values N = 85 (corresponding to a ration r_{i+1}/r_i = 1.10) and λ = 0.008. Scattering amplitudes $A(r_i,\theta_k)$ calculations have been performed with refractive index $n_s = 1.33$ and $n_p = 1.59$. The particle size distributions extracted from experimental data begin to show clear deviations compared to the expected ones if the analysis is performed with a 10 % error on the value of n_p/n_s.

Figure 3 shows the volume distributions of the various bimodal systems as calculated by our algorithm. As can be seen, the two populations have been resolved for every volume fraction investigated. The numerical values of the locations of the two peaks are overestimated typically by 10%, and the relative volume of these two distributions are always close to the input volume distribution even if the size ratio and thus the ratio of scattering amplitudes is large. However, in this case some spurious peaks are visible near the position of the large particles. This artifact may be attributed to the predominance of the scattering contribution from the large particles: for the mixture V(55):V(350) = 1:100, the expected scattered intensities range from I(55):I(350) = 1:2000 for $\theta = 30°$ to I(55):I(350) = 1:20 for $\theta = 120°$. The analysis of the four quasicontinuous systems is summarized in figure 4, which shows the cumulative volume (i.e $\int_0^r V(r')dr' / \int_0^\infty V(r')dr'$) as a function of the particle radius.

Although the different populations of monodisperse particles are not always separated (for example particles of radius 150 and 200nm are never separated), the general trend of these quasicontinuous distributions are always correctly reproduced. For a qualitative discussion, we characterize each volume distribution V(r) with it's mean value $\langle r \rangle = \int_0^\infty V(r)r dr$, the coefficient of variation $\sigma = \sqrt{m_2}/\langle r \rangle$, the skewness $m_3/m_2^{3/2}$, the kurtosis $m_4/m_2^2 - 3$, where $m_n = <(r - <r>)^n>$ is the nth moment of the volume distribution. Table 1. summarizes a comparison of these quantities as calculated for the experimental systems and obtained from the light scattering data. In every case these complex distributions are correctly reproduced.

Table 1: Analysis of quasi-continuous particle distributions

		$<r>$, nn	σ	Skewness	Kurtosis
System I	input	236	0.57	0.16	-1.36
	output	264	0.53	0.15	-1.62
System II	input	240	0.51	0.08	-1.07
	output	286	0.48	-0.02	-1.47
System III	input	316	0.34	-0.84	-0.03
	output	353	0.31	-1.18	-0.24
System IV	input	153	0.60	0.96	0.93
	output	181	0.64	1.34	0.84

The results obtained from the simultaneous multiangle analysis of experimental data were always better than results obtained from single-angle experiments performed at different angles. Table 2 shows the comparison of single- and multiple-angle analysis for a quasicontinuous particle size distribution. A single-angle analysis of a quasicontinuous system always misses some particle populations, and as a consequence, the characteristics of the particle size distribution differs significantly from the expected one. In every case, the simultaneous analysis of data collected at different angles increases significantly the realibility of the result.

Table 2: comparison of single- and multi-angle analysis for a quasi-continuous particle distribution

System I	$<r>$, nn	σ	Skewness	Kurtosis
Input	236	0.57	0.16	-1.36
Output 7 angles	264	0.53	0.15	-1.62
Output 30°	240	0.51	0.08	-1.07
Output 90°	286	0.48	-0.02	-1.47
Output 120°	316	0.34	-0.84	-0.03

Multi-angle Laplace inversion applied to turbid suspensions

The previous paragraph shows that we have been able to develop a simple method of analysis of multi-angle photon correlation spectroscopy data. The main advantage of this method is that the importance of the static scattered intensity measurement may be optimally matched with the characteristics of the optical set-up. The test experiments on a commercial instrument show that this method allows us to obtain quantitative information on the particle size distribution, despite the relatively short (15 min) acquisition time. If a priori information on particle size distribution is available, a further regularization may in principle be added in the inversion of the matrics system. Based on this method we can now try to apply it to PCS data obtained for turbid suspensions using a 3DDLS instrument.

We have prepared 3 different bimodal mixtures of polystyrene spheres with radii $r_1 = 57$ nm and $r_2 = 184$ nm and a mixing ratio of 10:1 by volume, where the total volume fraction varied between $\Phi = 0.011$ (sample 2), $\Phi = 0.0011$ (sample 3) and 0.00011 (sample 4). Sample 2 and 3 are already extremely turbid (see figure 5 A), which in the case of sample 2 yields a quite low value for the transport mean free path $l^* = 1.07$ mm at a wavelength of $\lambda = 680$ nm. Using standard autocorrelation measurements, a correct determination of the particle size distribution is barely possible for the most dilute sample. Already at an overall concentration of $\Phi = 0.0011$ (sample 3), the autocorrelation function is considerably shifted and contains a significant fast decay due to the contribution from multiple scattered light, and at the highest concentration (see figure 5) the multiple scattering is so strong that no measureable autocorrelation function can

16

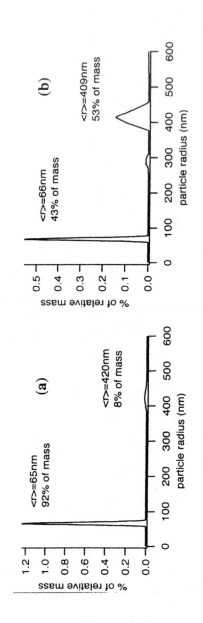

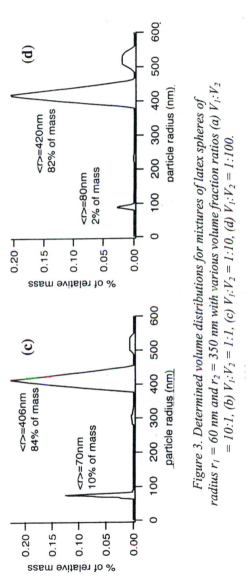

Figure 3. Determined volume distributions for mixtures of latex spheres of radius $r_1 = 60$ nm and $r_2 = 350$ nm with various volume fraction ratios (a) $V_1{:}V_2 = 10{:}1$, (b) $V_1{:}V_2 = 1{:}1$, (c) $V_1{:}V_2 = 1{:}10$, (d) $V_1{:}V_2 = 1{:}100$.

18

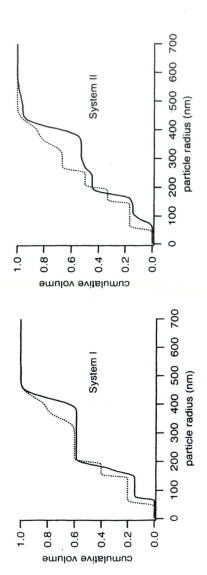

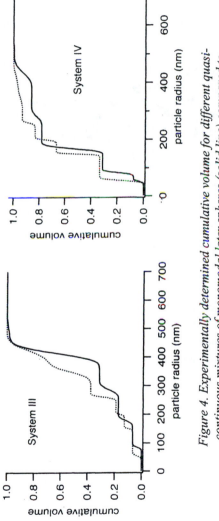

Figure 4. Experimentally determined cumulative volume for different quasi-continuous mixtures of monomodal latex spheres (solid line) compared to theoretically expected distributions (dotted line).

be obtained for a measurement period of 900 s at a scattering angle of θ = 90°
We note that in these experiments a diode laser with a rather short coherence
length was used and therefore the loss of coherence along the multiple scattering
paths, typical length $L^2/l^* \sim 10$ cm for sample 2, leads to a vanishing signal.

However, when using the 3D instrument to selectively detect single
scattering we obtain the correct correlation functions, although the intercept
decreases considerably at higher concentrations due to the non-negligeable
contribution from multiple scattering. Moreover, when combining multiangle
measurements using the 3D setup and simultaneous multiangle analysis, we can
recover the correct particle size distribution even at the highest particle
concentration (sample 2, see figure 5). It is important to note that even at the
lowest concentration a single angle analysis does not allow us to separate the
two populations and instead yields broad but homogeneous particle size
distributions. These examples clearly demonstrate that the combination of
3DDLS and multi-angle Laplace transformation can be used as a very powerfull
tool that extends the applicability of PCS to highly turbid samples.

Diffusing Wave Spectroscopy

Diffusing wave spectroscopy (DWS) extends standard PCS to media with
strong multiple scattering by treating the transport of light as a diffusion process
as first described by Maret and Wolf in 1987[14].

Method

The intensity autocorrelation function of the diffusely scattered light
$g_2(\tau) - 1 = \langle I(t)I(t+\tau)\rangle / \langle I\rangle^2 - 1$ can be expressed in terms of the mean square
displacement of the scattering particles

$$g_2(\tau) - 1 = \left[\int_0^\infty ds\, P(s) \exp\left(-(s/l^*)k_0^2 \langle \Delta r^2(\tau)\rangle\right)\right]^2 \qquad (15)$$

with $k_0 = 2\pi n/\lambda$ being the wave number of light in a medium with refractive
index n. P(s) is the distribution of photon trajectories of length s in the sample
and it can be calculated within the diffusion model taking into account the
experimental geometry. The transport mean free path l^* characterizes the typical
step length of the photon random walk, given by the individual particles
scattering properties and particle concentration (if necessary l^* can be
determined independently and enters the analysis as a constant parameter).
Therefore DWS is able to provide information about the size of particles in

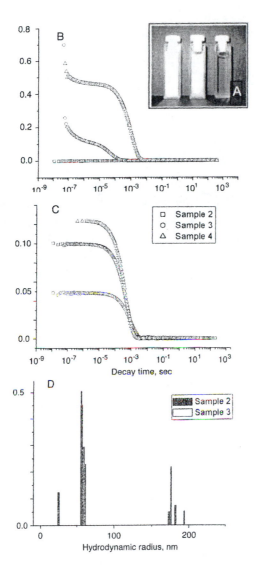

Figure 5. 3DDLS with turbid mixtures. (A) Sample 2 (left), 3 (middle), and a comparison with pure water (right). (B) Autocorrelation function of samples 2, 3 and 4 as measured at 90° with a measurement time of 600 s. The increased values at short decay times (< 10⁻⁶sec.) are due to detector afterpulsing. (C) 3D Cross-correlation function at 90° with a measurement time of 900 s for sample 2 and 3 and 300 s for sample 4. (E) Experimentally determined particle size distributions (by volume) from multi-angle 3DDLS..

dispersions without any restrictions on particle concentration and turbidity. Furthermore the DWS technique can be used to study and characterize a variety of complex systems such as dense colloidal dispersions and gels, emulsions, ceramic slurries and green bodies, biopolymers and gels (yoghurt and cheese), granular media, foams, polymers and concentrated surfactant solutions[17,30-37]. A typical setup for DWS measurements is shown in the Figure 6. For a large rectangular cell (layer thickness L >> l*, no absorption) the correlation function in reflection is given by

$$g_2(t) - 1 = \beta \exp\left[-a\sqrt{t}\right] = \beta \exp\left[-2\gamma\sqrt{\frac{6t}{\tau_0}}\right] \tag{16}$$

, where $D_0 = 1/\tau_0 k_0^2$ denotes the single particle short time diffusion constant and γ is a parameter of order l* taking account for the coupling of the incident light wave to the diffuse light in the sample. The backscattering geometry is particularily suited for particle sizing since the optical parameters of the sample, such as l* or the sample size, do not enter the analysis.

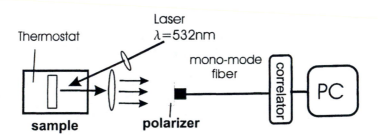

Figure 6. DWS-Setup for remote particle sizing. Fluctuation of the diffusively reflected light (incident laser with wavelength λ =532nm) are detected with a mono-mode fiber and subsequently analyzed with a digital correlator/PC. If possible the sample should be placed in a temperature controlled bath filled with an index matching fluid (e.g. water). A polarizer in front of the detector allows to distinguishbetween polarization preserving scattering (VV) or depolarized scattering (VH).

Particle Sizing with DWS

Plotting $\ln[g_2(t)-1]$ as a function of $\sqrt{t}$ results in a straight line of the measured data.

$$\ln[g_2(t)-1] \propto -a\sqrt{t} = -2\gamma\sqrt{6t/\tau_0} \qquad (17)$$

This expression that holds quite generally (even in the presence of absorption) for suffiently long times (but still $t < \tau_0$). From the linear slope the required information $a = 2\gamma\sqrt{6t/\tau_0}$ now can be easily obtained. If γ is known the (average) short time diffusion constant of Brownian particles is calculated from the decay time $\tau_0 = 1/Dk_0^2$. Subsequently the particle size is obtained using the Stokes-Einstein relation for non-interacting spheres: $D_0 = 1/k_0^2\tau_0 = kT/6\pi\eta R$. The main problem of this approach is that the value of γ is not constant but in fact can take values between 1 and 3 depending on particle size and detected polarization state (VH or VV as shown in the Figure 7)[38]. A way to overcome this problem is to carry out two measurements with orthogonal polarization states and subsequently average the obtained results as explained in Ref. [4].

$$R = k_0^2 \frac{kT}{6\pi\eta} \cdot \tau_0 \cong k_0^2 \frac{16kT}{\pi\eta} \cdot \left(\frac{\langle\gamma\rangle}{a_{VH} + a_{VV}}\right)^2 \qquad (18)$$

Absorption and limited container size (L<50l*)

Both absorption and limited container size can lead to a loss of photons along a given photon trajectory. In such cases the contribution of long paths is exponentially attenuated $P(s) = \exp(-s/l_a)P(s,l_a = \infty)$. This corresponds to a simple variable exchange $6t/\tau_0 \rightarrow 6t/\tau_0 + 3l^*/l_a$, where l_a is the "effective" absorption length (for non-absorbing particles in a infinite medium this is the absorption length of the pure solvent)[16]. This may strongly influence the decay half time $(g_2(t_{1/2})-1)/\beta = 1/2$ of the normalized correlation function which therefore is not a reliable paramter for the size analysis with DWS.

In the $\log[g_2(t)-1] \leftrightarrow \sqrt{t}$ representation however a finite l_a only leads to a rounding of the correlation function at short correlation times. The slope of this function at larger correlation times remains unaffected. Therefore the data should always be analyzed using this representation.

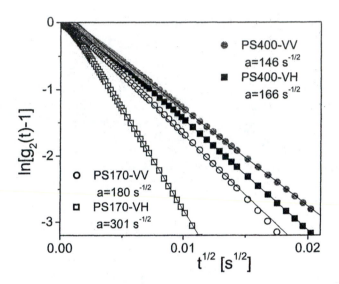

Figure 7. Particle Sizing with DWS: Monodisperse suspensions of polystyrene spheres at $\Phi\sim2\%$ vol. fract($T=22°$, 10 mm glass cuvettes).The measured correlation functions are shown for two examples detecting both polarized (VH) and depolarized (VV) reflected light. From the linear slope (fit: solid lines) in the $\log[g_2(t)-1] \leftrightarrow \sqrt{t}$ representation the particle size can be determined reliably. Example 1: Particle diameter d=2R=167nm [supplier: d= 170nm], Example 2: d=396nm [supplier: d= 400nm],

Polydispersity

DWS is not able to provide information about the size distribution. As a matter of fact light diffusion averages over the different particle sizes (Radius R) weighted by the particle scattering cross section, for moderate concentrations proportional to $1/l^*(R)$. Hence if $N(R)$ is the number distribution of radii the DWS effective radius is given by [4]:

$$1/R_{eff} = \int \frac{N(R)}{l^*(R)} \cdot \frac{1}{R} dR / \int \frac{N(R)}{l^*(R)} dR \tag{19}$$

The DWS effective radius R_{eff} is the apparent radius we expect from an analysis based on Eq.(18).

PCS Particle Sizing in Turbid Suspensions: An industrial application

The above development shall be illustrated by an application to particle sizing in suspensions of atrazin, an active agent used in crop protection. The initially submillimeter grains (from Novartis Crop Protection AG [now Syngenta Crop Protection AG], Münchwilen, Switzerland) were reduced in size by ultrasonic degradation (Jaroslav Ricka and René Nyffenegger, University of Berne, Switzerland). The suspensions contained 0.5 wt. % of the industrial surfactant soprophor (Novartis AG, Basel, CH) and were subsequently diluted with dionized water and soprophor at the same concentration (for some of the lower concentrations the soprophor content was 0.3% or even less which did however not affect the sample stability over the measurement time) .

Particle Size from traditional PCS and from 3-D DLS

Highly diluted samples (volume fraction $\Phi \leq 10^{-4}$) were characterized with a traditional PCS setup (Goniometer system from ALV GmbH, Langen – Germany, λ=488 nm).

Turbid suspensions where first analyzed with 3DDLS at $\Phi = 1\%$ (λ=633.2 nm). Higher concentrations could not be accessed with 3DDLS but only with DWS as shown below. The reason being that these high refractive index

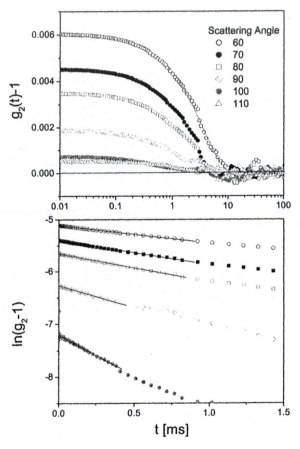

Figure 8. Praticle sizing of atrazin suspensions with 3DDLS. Measured correlation function at different scattering angles.

particles strongly scatter light. For the chosen concentration the scattering mean free path is as low as approx. 100 μm which means that the sample appears extremely turbid.

In Figure 8 we show the measured 3DDLS-correlation function for different scattering angles. For amplitudes as low as 2×10^{-3} the measurements give reliable results for the diffusion coefficient as shown in the next figure.

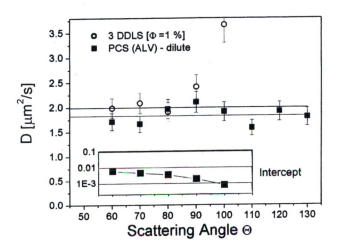

Figure 9. Particle diffusion coefficient from traditional PCS (highly diluted) and 3DDLS at $\Phi = 1\%$. $D = 1.9 \mu m^2/s$ *corresponds to a particle size of d=230nm.*

From Stokes-Einstein relationa a paticle diameter of $230 \pm 10\,nm$ is obtained from both traditional PCS [ALV-Goniometer], $\overline{D} = 1.83 \mu m^2/s$ and 3DDLS, $\overline{D} = 1.99 \mu m^2/s, \Theta \leq 80°$. It is also seen that at scattering angle larger than 90° the 3DDLS amplitude (intercept) drops sharply and for values below 10^{-3} the particle size determination was no longer possible due to the insufficient signal to noise ratio (Figure 9). The drop in amplitude can be explained by the fact that the particles predominantly scatter in forward direction whereas multiple scattering is more or less isotropic. The relative amount of single scattering therefore drops for large angles.

Our measurements illustrate how 3DDLS can be used to extend the range of application of 3DDLS to very turbid suspensions. In the limit of single scattering (dilute suspensions) 3DDLS reduces to traditional PCS. The performance is exactly the same as PCS but now also turbid suspensions can be characterized. The price to pay is that the measurement time increases as the amplitude (intercept) decreases. For the very small intercept values discussed here (of the order 10^{-3}) the measurement time chosen was ca. 4h for each scattering angle.

DWS Analysis at elevated densities

DWS offers an excellent alternative approachto characterize particle diffusion for such elevevated volume fractions and even above. We have analyzed the typical relaxation time τ_0 for a volume fraction up to $\Phi = 20\%$ as shown in Figure 10. For DWS there is no particular difficulty to access this regime. Measurements are straightforward and take of the order of minutes or less. For moderate concentrations (up to 2%) the time constant approaches the limit of free diffusion $\tau_0 = 1/Dk_0^2 \cong 1.8ms$ as before $(\lambda = 488nm)$.

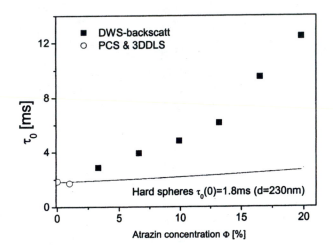

Figure 10. Relaxation time from DWS for dense suspensionsof atrazin. At low volume fractions the result from single scattering [PCS&3DDLS] are close. With increasing concentration interparticle interactions lead to a slowed particle diffusion and consequently to an increase in τ_0. Solid line: Expected increase for hard-spheres.

At higher volume fractions the relaxation time strongly increases. Also shown is the expected increases for the case of hard spheres. However, the measured concentration dependence is found to be significantly more pronounced. One reason might be the anisotropy of the particles slowing down diffusion already at moderate densities.

PCS Particle Sizing in Turbid Suspension: Scope and Limitations

With the advent of new technologies in PCS, such as 3DDLS and DWS, the characterization of turbid suspensions is a straightforward task if the necessary equipment is available. The information typically required (the distribution of diffusive relaxation times) can be obtained from 3DDLS at concentrations between 1% and 30% (volume fraction) depending on the refractive index of the particles (requirement: scattering mean free path $l > 50\mu m$). For turbid samples DWS gives access to the diffusive particle relaxation time and the mean particle size. In the range $l \sim 50\text{-}500$ μm both methods can be applied.

While 3DDLS provides very precise information it also may require long measurement times, depending on the sample turbidity. DWS experiments however can be done in minutes, independent of turbidity. Additionally the DWS-backscattering geometry allows remote sensing or the construction of flexible sensors. But the information obtained from DWS is smeared out by the multiple scattering of light and particle size distributions cannot be accessed.

Problems appear in both cases when the relaxation time (or diffusion constant) is altered by particle-particle interactions and increased suspension viscosity. In this case the Stokes-Einstein relation in its simple form is no longer applicable for the size analysis. Nevertheless a lot of information is available from such measurements. This requires however a more detailed analysis of the results and the particle radius cannot be extracted in a simple fashion.

In general both 3DDLS and DWS can be applied safely for volume contens less than 5% while for higher volume fractions some information about the sample properties is needed or results at different concentrations should be compared to check for consistency.

Acknowledgements

The original atrazin samples were provided by Novartis CP, Münchwilen (CH). We thank A. Schofield for providing the PMMA samples. We thank R. Hilfiker, M.v. Raumer, H-W. Haesslin, I. Flammer, R. Nyffenegger and J. Ricka for discussions. This work was supported by the Swiss National Science Foundation, TOP NANO 21 and the Novartis Technology Board.

References

1. Schurtenberger, P. & Newman, M. E. in *Environmental Particles* (eds. Buffle, J. & van Leeuwen, H.) 37-115 (Lewis Publishers, Boca Raton, 1993).
2. Pusey, P. (ed. Zemb, Lindner) (North Holland, Elsevier, Amsterdam, 2002).
3. Pusey, P. N. & Van Megen, W. Detection of Small Polydispersities by Photon-Correlation Spectroscopy. *Journal of Chemical Physics* **80**, 3513-3520 (1984).
4. Bowen, P. Particle size distribution measurement from millimeters to nanometers, and from rods to platelets. *Journal of Dispersion Science and Technology* **23**, 591-599 (2002).
5. Segre, P. N., Vanmegen, W., Pusey, P. N., Schatzel, K. & Peters, W. 2-Color Dynamic Light-Scattering. *Journal of Modern Optics* **42**, 1929-1952 (1995).
6. Pusey, P. N. Suppression of multiple scattering by photon cross-correlation techniques. *Current Opinion in Colloid & Interface Science* **4**, 177-185 (1999).
7. Schatzel, K. Suppression of Multiple-Scattering by Photon Cross-Correlation Techniques. *Journal of Modern Optics* **38**, 1849-1865 (1991).
8. Urban, C., Romer, S., Scheffold, F. & Schurtenberger, P. Static and dynamic light scattering in turbid suspensions. *Macromolecular Symposia* **162**, 235-247 (2000).
9. Urban, C. & Schurtenberger, P. Characterization of turbid colloidal suspensions using light scattering techniques combined with cross-correlation methods. *Journal of Colloid and Interface Science* **207**, 150-158 (1998).
10. Urban, C. & Schurtenberger, P. Application of a new light scattering technique to avoid the influence of dilution in light scattering experiments with milk. *Physical Chemistry Chemical Physics* **1**, 3911-3915 (1999).
11. Aberle, L. B., Hulstede, P., Wiegand, S., Schroer, W. & Staude, W. Effective suppression of multiply scattered light in static and dynamic light scattering. *Applied Optics* **37**, 6511-6524 (1998).
12. Overbeck, E., Sinn, C., Palberg, T. & Schatzel, K. Probing dynamics of dense suspensions: Three-dimensional cross-correlation technique.

Colloids and Surfaces a-Physicochemical and Engineering Aspects **122**, 83-87 (1997).

13. Maret, G. Diffusing-wave spectroscopy. *Current Opinion in Colloid & Interface Science* **2**, 251-257 (1997).

14. Maret, G. & Wolf, P. E. Multiple Light-Scattering from Disordered Media - the Effect of Brownian-Motion of Scatterers. *Zeitschrift Fur Physik B-Condensed Matter* **65**, 409-413 (1987).

15. Pine, D. J., Weitz, D. A., Chaikin, P. M. & Herbolzheimer, E. Diffusing-Wave Spectroscopy. *Physical Review Letters* **60**, 1134-1137 (1988).

16. Weitz, D. A., Zhu, J. X., Durian, D. J., Gang, H. & Pine, D. J. Diffusing-Wave Spectroscopy - the Technique and Some Applications. *Physica Scripta* **T49B**, 610-621 (1993).

17. Rojas-Ochoa, L. F., Romer, S., Scheffold, F. & Schurtenberger, P. Diffusing wave spectroscopy and small-angle neutron scattering from concentrated colloidal suspensions. *Physical Review E* **65**, art. no.-051403 (2002).

18. Berne, B. J. & Pecora, R. *Dynamic Light Scattering* (John Wileys and Sons, New York, 1976).

19. Gisler, T. et al. Mode-Selective Dynamic Light-Scattering - Theory Versus Experimental Realization. *Applied Optics* **34**, 3546-3553 (1995).

20. Finsy, R., De Groen, P., Deriemaeker, L., Gelade, E. & Joosten, J. Data Analysis of Multi-Angle Photon Correlation Measurements without and with Prior Knowledge. *Part. Part. Syst. Charact.* **9**, 237-251 (1992).

21. Cummins, P. G., Staples, E. J. & Langmuir. Particle Size Distributions Determined by a "Multiangle" Analysis of Photon Correlation Spectroscopic Data. *Langmuir* **3**, 1109-1113 (1987).

22. Bryant, G. & Thomas, J. C. Improved Particle Size Distribution Measurements Using Multiangle Dynamic Light Scattering. *Langmuir* **11**, 2480-2485 (1995).

23. De Vos, C., Deriemaeker, L. & Finsy, R. Quantitative Assessment of the Conditioning of the Inversion of Quasi-Elastic and Static Light Scattering Data for Particle Size Distributions. *Langmuir* **12**, 2630-2636 (1996).

24. Wu, C., Unterforsthuber, K., Lilge, D., Luddecke, E. & Horn, D. Determination of Particle Size Distribution by the Analysis of Intensity-Constrained Multi-Angle Photon Correlation Spectroscopic Data. *Part. Part. Syst. Charact.* **11**, 145-149 (1994).

25. Provencher, S. W. Inverse Problems in Polymer Characterization : Direct Analysis of Polydispersity with Photon Correlation Spectroscopy. *Makromolekulare Chemie-Macromolecular Chemistry and Physics* **180**, 201-209 (1979).

26. Lawson, C. L. & Hanson, R. J. *Solving least squares problems* (Prentice-Hall, Englewood Cliffs (N.J.), 1974).

27. Barber, P. W. & Hill, S. C. *Light Scattering by Particles: Computational Methods* (World Scientific, Singapore, 1990).

28. Ostrowsky, N., Sornette, D., Parker, P. & Pike, E. R. Exponential sampling method for light scattering polydispersity analysis. *Optic. Acta* **28**, 1059-1070 (1981).

29. Glatter, O. & Hofer, M. Interpretation of static light scattering data.III Determination of Size Distributions of Polydisperse Systems. *Journal of Colloid and Interface Science* **122**, 496-506 (1988).

30. Cardinaux, F., Cipelletti, L., Scheffold, F. & Schurtenberger, P. Microrheology of giant-micelle solutions. *Europhysics Letters* **57**, 738-744 (2002).

31. Romer, S. et al. Simultaneous light and small-angle neutron scattering on aggregating concentrated colloidal suspensions. *Journal of Applied Crystallography* **36**, 1-6 (2003).

32. Romer, S., Scheffold, F. & Schurtenberger, P. Sol-gel transition of concentrated colloidal suspensions. *Physical Review Letters* **85**, 4980-4983 (2000).

33. Wyss, H. M., Romer, S., Scheffold, F., Schurtenberger, P. & Gauckler, L. J. Diffusing-wave spectroscopy of concentrated alumina suspensions during gelation. *Journal of Colloid and Interface Science* **241**, 89-97 (2001).

34. Schurtenberger, P., Stradner, A., Romer, S., Urban, C. & Scheffold, F. Aggregation and gel formation in biopolymer solutions. *Chimia* **55**, 155-159 (2001).

35. Rojas, L. F. et al. Particle dynamics in concentrated colloidal suspensions. *Faraday Discussions* **123**, 385-400 (2003).

36. Hebraud, P., Lequeux, F., Munch, J. P. & Pine, D. J. Yielding and rearrangements in disordered emulsions. *Physical Review Letters* **78**, 4657-4660 (1997).

37. Gang, H., Krall, A. H. & Weitz, D. A. Thermal Fluctuations of the Shapes of Droplets in Dense and Compressed Emulsions. *Physical Review E* **52**, 6289-6302 (1995).

38. Mackintosh, F. C., Zhu, J. X., Pine, D. J. & Weitz, D. A. Polarization Memory of Multiply Scattered-Light. *Physical Review B* **40**, 9342-9345 (1989).

Chapter 2

Multiple Light Scattering Methods for Dispersion Characterization and Control of Particulate Processes

P. Snabre[1], L. Brunel[2], and G. Meunier[2]

[1]Centre de Recherche Paul-Pascal, CNRS, Avenue Albert Schweitzer,
33600 Pessac, France
[2]Formulaction, 10 Impasse Borde Basse, 31240 L'Union, Toulouse, France

This work concerns multiple light scattering methods for the characterization of concentrated dispersions. After recall of the wave transport theory and some experimental means for analyzing incoherent wave propagation in heterogeneous media, the present paper concerns a new imagery method for the optical characterization of concentrated dispersions and applications in the field of industrial processes.

Introduction

Control of microstructure and dispersion stability in industrial reactors requires new investigation methods for estimating average scatterer size and particle volume fraction. Optical extinction sensors only concern sufficiently diluted systems and sizing based on reflexion probes remains inaccurate unless the detector is calibrated against the material under test (1). In the multiple scattering regime, spatial or time distribution of wave energy density of multiply scattered from a dense collection of particles is representative of dispersion microstructure. After recall of the wave transport theory and experimental methods for measuring photon transport mean free path in random scattering media, the first part describes a less common imagery method based on two dimensional analysis of wave energy density in the incoherent backscattered spot light for dispersion characterisation. The analysis of the microstructure of superficial molten glass foam in hot temperature glass tank furnaces illustrates the usefluness of the imagery method for dispersion characterization. The second part concerns dispersion characterization with the optical analyzer Turbiscan commercialised by the society Formulaction and the last section presents some applications in the field of industrial processes.

Light Transport in Concentrated Dispersions

Scattering and Transport Mean Free Paths

One considers a plane-wave (wave vector k) incident on a random isotropic collection of particles (average diameter d, particle volume fraction ϕ). In the weak scattering regime or Born approximation ($kl > 1$), the scattering mean free path $l(kd,\phi)$ determines the exponential extinction of the coherent field through the medium (2):

$$\frac{1}{l(kd,\phi)} = \frac{N}{k^2} \int_0^{2k} f(kd,q)\, S(kd,\phi,q)\, q\, dq \tag{1}$$

where the form factor $f(kd,q)$ describes single particle properties depending upon scattering angle through the scattering wave vector q while the structure factor $S(kd,\phi,q)$ accounts for space correlations among scatterers. In the diluted regime ($\phi < 0.01$), the inverse $1/l$ of the scattering mean free path scales as the particle volume number N as a result of independent scattering from particles (structure factor $S = 1$).

In the photon diffusion approximation, the propagation of multiply scattered waves is assumed incoherent. Statistics of incoherent scattering paths then can be described in terms of photon random walks. A renormalization of anisotropic random walks makes appear the transport mean free path $l^* = l / (1-g)$ that scales as the scattering mean free path l and depends upon the asymmetry factor g defined in terms of the average cosine of scattering angles (Figure 1) (3).

In the case of isotropic scatterers (or Rayleigh scatterers with $kd \ll 1$), scattering and transport mean free paths are identical (Figure 2). For large particles ($kd > 1$) mainly scattering in the forward direction, the transport mean free path l^* becomes larger than the scattering mean free path l (Figure 2).

From statistical arguments, one can derive the mean quadratic displacement $<\rho^2> \approx n \, l \, l^*$ of n scattering paths from the injection point (Figure 1) and define a photon diffusion coefficient $D \approx c \, l^*$ scaling as the product of light speed c and transport mean free path l^*.

In the case of non-conservative scattering, transport mean free path l^* and absorption length l_a then determine the incoherent propagation of multiply scattered waves in relation with dispersion microstructure.

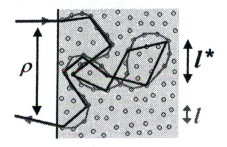

Figure 1. *Renormalization of anisotropic diffusion paths. Scattering mean free path l, transport mean free path l^* and photon displacement ρ from the injection point*

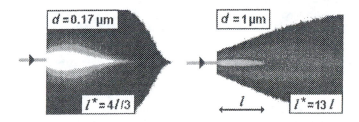

Figure 2. *Coherent extinction (yellow and red areas) of He-Ne laser light and incoherent propagation of scattered waves (blue area) in diluted latex beads suspension.*
Left figure : l=6mm, l* =8mm for d=0.17µm and $\phi = 5\,10^{-4}$.
Right figure : l=6mm, l* =78mm for d=1µm and $\phi = 5\,10^{-3}$.

Incoherent Backscattered Spot Light.

The spectral transport mean free path l^* is usually derived from diffuse reflectance measurements (4), the temporal spread of an ultrashort light pulse (5), the phase shift of diffuse amplitude - modulated photon density wave (6) or the angular analysis of the coherent backscattering cone (7).

A less common method is based on the two dimensional analysis of the light density energy in the incoherent backscattered spot light when illuminating the scattering medium with a point light source (3). For a point light source normally incident on a non absorbing heterogeneous medium, the transport mean free path l^* indeed determines the space distribution of wave energy density in the incoherent backscattered spot light.

For long diffusion paths (exit distance $\rho \gg l^*$), the surface energy density F derived either from Green functions formalism or Monte Carlo simulations scales as the transport length l^* and the inverse cube of radial distance as a result of the diffusive nature of wave propagation. On the other hand, the radial distribution of short diffusion paths no longer obeys a Gaussian statistics and displays a weaker radial scaling exponent of about 1.4. (3):

$$F(\rho) = \frac{1}{\pi\, l^{*2}} \left(\frac{l^*}{\rho} \right)^3 e^{-(\rho/l^*)\sqrt{l/l_a}} \qquad \text{for } \rho > 4l^* \qquad (2)$$

$$F(\rho) \approx \frac{1}{\pi\, l^{*2}} \left(\frac{4l^*}{\rho} \right)^{1.4} e^{-(\rho/l^*)\sqrt{l/l_a}} \qquad \text{for } \rho < 4l^* \qquad (3)$$

where the exponential term accounts for absorption phenomena (3). The sharp radial decrease of the surface energy density F over more than 6 decades indeed shows a center part relative to short path photons and a peripheral region for long path photons with a scaling of the light flux as the inverse cube of radial distance (Figure 3).

Imagery analysis of the diffuse spot light with a high dynamic camera then provides a powerful way to derive the transport mean free path l^* and estimate the average size of scatterers. For a polydisperse system, scattering from different size particles results in diffusion paths with different step sizes. In the diffusion approximation, the statistics of random walks still reduce to Gaussian functions with a single average step size $<1/l>^{-1}$ weighted by the form factor $f(kd,q)$ and the structure factor $S(kd,\phi,q)$ of scattering species. The space distribution of wave energy density in the diffuse spot light then involves the average transport length $< 1 / l^* >^{-1}$ and cannot yield information about polydispersity. For large non interacting particules ($f \approx k^2d^2$ and $S=1$), the

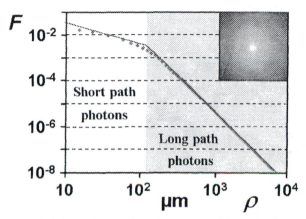

Figure 3. *Radial dependence of surface energy density in the incoherent backscattered spot light (see insert) for a non-absorbing latex suspension.*

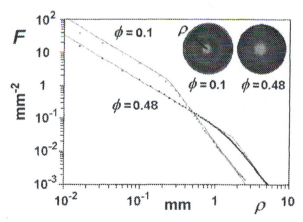

Figure 4. *Radial dependence of surface energy density in the incoherent backscattered spot light. Experiments with 0.17μm latex particles in water and model predictions ($l^* = 66μm$ for $\phi = 0.1$ and $l^* = 430$ μm for $\phi = 0.48$).*

average transport length scales linearly with the inverse of particle area per unit volume.

For monodispersed latex beads suspensions, the diameter dependence of the transport mean free path well agrees with predictions from Mie – Percus Yevick (MPY) theory for hard spheres in purely hydrodynamic interactions (Figure 4).

However, the Percus Yevick approximation underestimates the transport mean free path l^* for highly concentrated 0.17μm latex beads in water because of non negligible repulsive interactions and formation of crystallites as shown by microscopic observations (Figure 5). Experiments performed in a Couette flow indeed indicate hydrodynamic melting of crystallites in relation with a decrease of the transport mean free path at shear rates $\dot{\gamma} > 10s^{-1}$ (Figure 5).

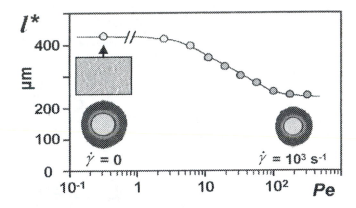

Figure 5. *Transport mean free path l^* versus Peclet number $Pe = \eta d^3 \dot{\gamma}/8kT$*
for 48% suspension of 0.17μm latex spheres in a Couette flow
(η is the suspension viscosity, $\dot{\gamma}$ the shear rate and kT the thermal energy)
Microphotography of the dispersion at rest and visualization
of the diffuse spot light at shear rates $\dot{\gamma} = 0$ and $\dot{\gamma} = 10^3 s^{-1}$.

The analysis of the light density energy in the incoherent diffuse spot light for investigating the microstructure of superficial molten glass foam in hot temperature glass tank furnaces illustrates the usefulness of the imagery method.

A pulsed Neodime Yag laser was used to visualize the diffuse light spot on superficial high temperature wet foam in a glass tank furnace (Figure 6). In the framework of optical geometry approximations, the measured transport mean free path $l^* \approx 5mm$ is representative of 1mm diameter air bubbles and was useful to describe the radiative transfer and temperature field in molten glass bulk.

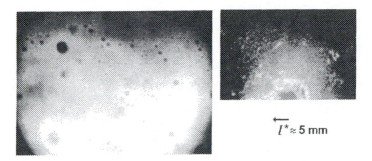

$l^* \approx 5$ mm

Figure 6. *High temperature superficial foam in a glass tank furnace and visualization of the diffuse spot light of characteristic size l* ≈ 5mm.*

Dispersion Characterization with Turbiscan Analyser

The Turbiscan optical analyser performs measurements of either transmission or diffuse reflectance from a turbid medium to analyse dispersion stability and estimate the average diameter of scatterers.

Turbiscan Analyser

The optical analyzer Turbiscan commercialized by the society Formulaction (Toulouse, France) uses a pulsed near infra red light source and detects light flux either transmitted or multiply backscattered from a dispersion. Two synchronous photodetectors move along the vertical tube containing the suspension and analyze both transmission and backscattering levels every 40μm along the cell height (Figure 7)

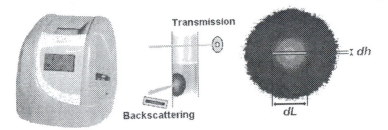

Figure 7. *Turbiscan device performing transmission and backscattering level measurements along the cell height.*

The diffuse spot light of characteristic size l^* is observed through a thin detection slot of thickness dh. From the photon diffusion approximation and for conservative scattering, the light flux through the detection area or diffuse reflectance R only depends on the dimensionless ratio l^* over dh. The diffuse reflectance R indeed scales as $(l^*/dh)^{1/2}$ in agreement with experimental observations (Figure 8) and theoretical considerations (8) independently of the scattering behavior of non absorbing species.

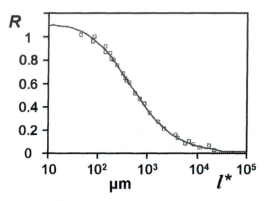

Figure 8. *Diffuse reflectance R measured with Turbiscan device versus transport mean free path* l^* *for latex beads suspensions, silicone and oil in water emulsions (the average diameter of nearly monodispersed particles or droplets ranges from 0.05μm up to 20μm).*

Concentrated Latex Beads Suspensions

Monodispersed latex beads suspensions were analysed with Turbiscan optical analyser. Figure 9 shows the particle volume fraction dependence of both transmission T and diffuse reflectance R for 0.17μm latex spheres in water. Transmission data are considered in the diluted regime while backscattering data are used in the multiple scattering regime for the analysis of the dispersion stability.

From the transmission T or the diffuse reflectance R, the extinction mean free path l or the transport mean free path l^* are derived and then average particle diameter or particle volume fraction can be estimated from the MPY approximation for hard spheres. Such an approximation only holds if physico chemical interactions between particles act over length scales much smaller than the average distance between particle surfaces.

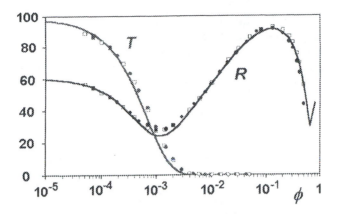

Figure 9. *Transmission level T and backscattering level R versus particle volume fraction for 0.17μm latex spheres in water. Experiments and model predictions (particle refractive index n_p=1.59).*

Control of Industrial Processes

Ultrasound Emulsification Process.

Ultrasound emulsification of 25% kerosene oil in water was studied with on line Turbiscan (Figure 10). Figure 11 displays time evolution of the backscattering level R during batch emulsification of 25% kerosene oil in water dispersion flowing in an optical cell.

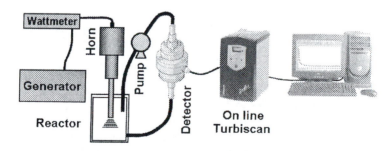

Figure 10. *Schematic drawing of on line Turbiscan device for analysis of batch ultrasound emulsification.*

The measured diffuse reflectance R increases with time and ultrasound power as a result of droplet break up. The equilibrium average droplet diameter derived from the average transport mean free path shows weak dependence upon ultrasound power in agreement with data from Mastersizer device (Malvern) after dilution of the emulsion (Figure 11) *(9)*. On line Turbiscan provides an efficient way to control both kinetics and efficiency of ultrasound emulsification in concentrated emulsions .

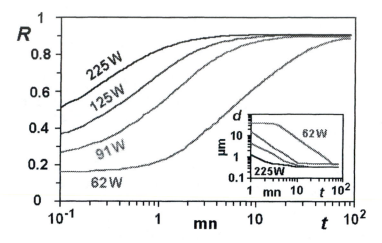

Figure 11. Time evolution of the backscattering level R and average droplet diameter d (see insert) as a function of ultrasound power during batch emulsification of 25% kerosene oil in water (9).

Calcium Carbonate Wet Grinding

Wet grinding of calcium carbonate particles was investigated with on line Turbiscan. The calcium carbonate suspension flowing out the mill is analyzed in the optical cell of on line Turbiscan (Figure 12) *(10)*.

The diffuse reflectance R of 4% calcium carbonate suspension in water increases with time as a result of particle grinding (Figure 13). However, grinding efficiency decreases with feed flow rate as particle mean residence time in the reactor is reduced. Equilibrium average particle diameter either obtained from Turbiscan analyzer or Mastersizer device are rather close as shown in the insert (Figure 13). On line Turbiscan is well suited for studying transient phenomena in grinding processes.

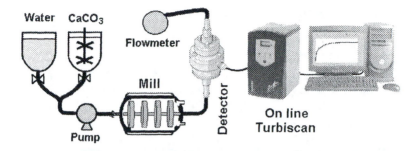

Figure 12. *Schematic drawing of on line Turbiscan device for analysis of batch grinding of calcium carbonate particles in water.*

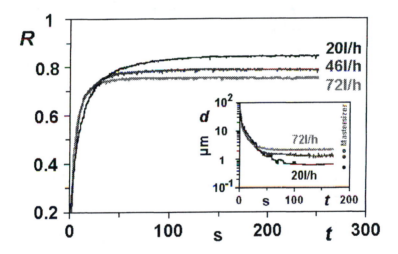

Figure 13. *Time evolution of the backscattering level R and average particle diameter d (see insert) as a function of feed flow rate during grinding of 4% calcium carbonate dispersion (10).*

In conclusion, the analysis of the backscattered spot light with a two dimensional (imagery device) or one dimensional detector (Turbiscan device) provides a quantitative way to investigate dispersion microstructure and control industrial processes.

References

1. Hartge, E.U.; Rensner, D.; Werther, *J. Chem. Ing. Tech.* **1989**, *61*, 1781.
2. Ishimaru, A. and Kuga, K.; *J. Opt. Soc. Am.* **1982**, *72*, 1317.
3. Snabre, P. and Arhaliass, A., *Appl. Opt.* **1998**, *37(18)*, 4017.
4. Kaplan, P.D.; Dinsmore, A.D.; Yodh, A.G. and Pine, D.J., *Phys. Rev. E.* **1994**, *50(6)*, 4827.
5. Bruce, N.E. and French P.M.W., *Appl. Opt.* **1995**, *34(25)*, 5823.
6. Banerjee, S.; Shinde, R and Sevick, E.M, *J.C.I.S.* **1999**, 209, 142.
7. Akkermans, E.; Wolf, P.E. and Maynard R., *Phy. Rev. Lett.* **1986**, *56*, 1471.
8. Snabre, P.; Mengual, O. and Meunier, G, *Coll. & Surf. A..* **1999**, *152*, 79.
9. Abismail, B; Delmas, H, *Ultrasonics Sonochem.* **1999**, *6*, 75.
10. Bordes, C; Garcia, F; Snabre, P; Frances, C, *Powder Technology*, **2002**, *128*, 2.

Chapter 3

Particle Size and Rapid Stability Analyses of Concentrated Dispersions: Use of Multiple Light Scattering Technique

P. Bru[1], L. Brunel[1], H. Buron[1,*], I. Cayré[1], X. Ducarre[1], A. Fraux[1], O. Mengual[1], G. Meunier[1], A. de Sainte Marie[1], and P. Snabre[2]

[1]Formulaction, 10 Impasse Borde Basse, 31240 L'Union, Toulouse, France
[2]Centre de Recherche Paul Pascal, Avenue Albert Schweitzer, 33600 Pessac, Bordeaux, France

Characterization of colloidal systems and investigation of their stability in their native state (*i.e.* without denaturation) is of prime importance for the formulator who wants to optimize the development of new products. A new technique has been developed, based on Multiple Light Scattering (MLS) to measure and analyze instability phenomena in liquid colloidal dispersions from 0 to 95% in volume fraction, with particles from $0.1\,\mu m$ to $1\,mm$, 5 to 50 times quicker than the naked eye. It is also a useful technique to characterize the dispersion state of colloidal samples (for quality control purposes) and the mean diameter of particles in dispersions (for analytical purposes).

Figures 1, 2,5–12, and 14 can be found in the color insert following page 52.

Introduction

Colloidal dispersions are thermodynamically unstable systems undergoing various types of instability phenomena that can be classified in two types: particle migration, which is due to differences in density between the dispersed and the continuous phases and includes creaming and sedimentation phenomena; and particle size variation, which is due to modifications of the size of the particles being reversible (flocculation) or irreversible (coagulation, coalescence, Ostwald ripening). Such instability phenomena can be more or less fast and their measurement is often left to visual inspection of the samples. However, the naked eye is quite subjective and can become rapidly tedious. The Multiple Light Scattering (MLS) technique associated to a vertical scanning of the sample is like a high resolution electronic eye, enabling to visualize instability phenomena before they are visible to the eye and to compute physical parameters and kinetics in order to facilitate comparisons between formulations. Moreover, the analysis is objective and reliable, as it does not depend on the operator.

Typical light scattering techniques used to characterize colloidal dispersions require the dilution of the sample as they are based on single diffusion models. However, dilution can be an issue regarding the integrity of the product as phenomena such as coalescence, de-flocculation, aggregation, *etc.* can occur due to the severe dilution taking place (1000 fold in most cases). Therefore, a technique that does not require any dilution nor denaturation of the sample is more appropriate to understand the actual phenomena and the dispersion state of the system. The MLS technique is offering this possibility and enables to measure the photon transport mean free path l^*, which is directly related to the ratio diameter/volume fraction. When associated to a vertical scanning of the sample, it is possible to get the spatial distribution of l^*. This value is a real fingerprint of the product as it gives information on the homogeneity of the sample and is characteristic of the dispersion. Therefore, it is a simple and useful tool for quality control purposes. Moreover, using the theories of Mie, Rayleigh and general optics, it is possible to derive the mean diameter of the particles. Because this value is obtained without any dilution, it enables to see the real state of the particles in the system.

In this chapter, we firstly describe the MLS technique and how it is used in the Turbiscan® technology and secondly we give examples of applications in stability measurements and characterization of colloidal systems.

Basics on Multiple Light Scattering

Definitions

We consider a suspension of non-absorbing spheres randomly dispersed in a transparent fluid and illuminated with a narrow light beam. In a first approximation, we neglect interferences between waves scattered from particles (independent scattering regime).

In these conditions the mean free path of the light, l, can be estimated through the particle surface density (derived from particle mean diameter d and particle volume fraction ϕ) and the scattering efficiency factor Q_s (1):

$$l(\phi,d) = \frac{1}{n\,(\pi d^2/4)\,Q_s} = \frac{2\,d}{3\,\phi\,Q_s} \quad \text{and} \quad \phi = n\,\frac{\pi\,d^3}{6} \tag{1}$$

where n is the particle density.

For Mie scatterers and larger ones ($\lambda/10$ up to 10λ, λ being the wavelength of the incident light), light scattering becomes anisotropic. This anisotropy can be characterised by the asymmetry factor g, which is the average cosine $<\cos\theta>$ of the scattering angles weighted by the phase function or scattering diagram $P(\theta)$ of the scatterer ($g = 0$ for isotropic Rayleigh scatterers and $0 < g < 1$ for Mie scatterers) (1,2). For non-isotropic scatterers, we further define the photon transport mean free path $l^* = l/(1-g)$ representing a decorrelation length above which the photon "forgets" the direction of the incident beam (2-4).

If the distance between particles becomes shorter than the wavelength (for small particles and/or high particle volume fraction), light scattering is no more independent, which increases both photon mean free path l and transport length l^* (3). This phenomenon is also included in the models as a second approximation.

Backscattering and Transmission Theory

Backscattering Physical Model

When a narrow light beam propagates into an optically thick dispersion contained in a glass measurement cell (a cylindrical cell here), the backscattered light spot shown in Figure 1 displays two regions:

× A central part corresponding to short path photons, which undergo a few scattering events before escaping the medium.

× A peripheral part corresponding to long path photons, which undergo a large number of scattering events before escaping the medium.

The characteristic size of the backscattered spot light is representative of the **photon transport mean free path** l^* (Figure2). The backscattered light flux BS measured through a thin detection area of thickness dh (Figure 1) scales as $(dh/l^*)^{1/2}$ in agreement with experimental observations (6):

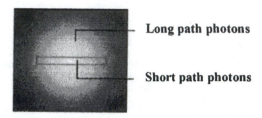

Long path photons

Short path photons

Figure1. Backscattered light spot

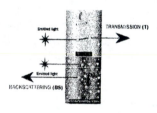

Figure 2. Representation of l and l*

$$BS \approx \sqrt{\frac{dh}{l^*}} \qquad (2)$$

Using the Mie theory corrected for high volume fraction with the approximation of Percus Yevik *(3)*, the transport mean free path l^* scales as particle mean diameter and the inverse of particle volume fraction:

$$l^*(d,\phi) = \frac{2d}{3\phi(1-g)Qs} \qquad (3)$$

where the asymmetry factor g and the scattering efficiency Q_s are derived from Mie theory.

Transmission Physical Model

The **photon mean free path** l (Figure 2) represents the mean distance traveled by photons before undergoing a scattering phenomenon. The Lambert - Beer law gives an analytical expression of the transmission T, measured by the optical analyzer as a function of the photon mean free path l:

$$T(l,r_i) = T_0 \ e^{-\frac{2\,r_i}{l}} = T_0 \ e^{-\frac{3\,r_i}{d}\,\phi Qs} \qquad (4)$$

where r_i is the measurement cell internal radius and $T_0(n_f)$ the transmission for the continuous phase. Therefore, the transmission T directly depends on the particle mean diameter d and particle volume fraction ϕ.

Effect of the Particle Diameter

Figure 3 shows, as an example, the backscattering level BS (experimental values and models from Mie) versus mean diameter d for latex particles for a constant volume fraction $\phi = 1\%$.

In accordance to the predictions of the physical model, experiments show an increase of the backscattered light flux with particle mean diameter for scatterers smaller than the incident wavelength (880 nm) and a decrease with particle size for large scatterers.

Effect of the Particle Volume Fraction

Figure 4 shows, as an example, the backscattering and the transmitted flux (experimental values and models from Mie theory) versus the particle volume fraction ϕ for a latex beads suspension with a constant particle diameter of 0.3 μm.

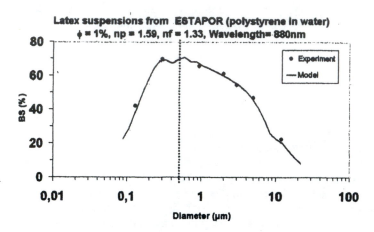

Figure 3. *Backscattering level BS versus particle mean diameter for latex particles in water.*

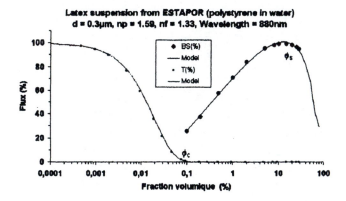

Figure 4. *Backscattering level BS and transmission T versus particles volume fraction for latex suspensions of 0.3μm in water.*

In the concentrated regime ($\phi > \phi_c$), both the transport model and the experiments show an increase of the backscattering level BS with particle volume fraction before reaching a maximum for a volume fraction ϕ_s. The critical volume fraction ϕ_c between diluted and concentrated regimes corresponds to a photon transport mean free path $l^*(\phi_c)$ equals to the measurement cell diameter $2r_i$.

In the diluted regime ($\phi < \phi_c$), the transmission T decreases exponentially with particle volume fraction in good accordance with the physical model and reaches a zero value in the concentrated regime ($\phi > \phi_c$).

The saturation of the backscattering signal for highly concentrated suspensions ($\phi > \phi_s$) results from dependent scattering effects. Indeed, for high volume fraction, scatterers display spatial organization leading to light interferences between waves scattered from particles. Such interferences induce an increase of the transport mean free path l^*, and reduce the backscattered light flux.

Stability Analysis and Characterization

Principle of the Measurement

The central part of the optical scanning analyser, Turbiscan®, is a detection head, which moves up and down along a flat-bottomed cylindrical glass cell (Figure 5). The detection head is composed of a pulsed near infrared light source ($\lambda = 880$ nm) and two synchronous detectors. The transmission detector (at 180°) receives the light, which goes through the sample, while the backscattering detector (at 45°) receives the light backscattered by the sample. The detection head scans the entire height of the sample (55 mm), acquiring transmission and backscattering data every 40 μm. It can also be used in "fixed position" mode where the head is set at a fixed sample height and can make acquisitions every 0.1 seconds. This latter mode is of particular interest for monitoring very quick instability phenomena such as breaking of foam.

The Turbiscan® (Figure 6) makes scans at various pre-programmed times and superimposes the profiles on one graph in order to show the destabilisation. Backscattering and/or transmission fluxes are shown in ordinate and the height of the cell in abscissa (Figure 7).

Stability Analysis

The optical analyser enables to monitor and quantify instability phenomena before being visible to the eye. In this section, we give examples of the typical instabilities and how they are detected and analysed by the Turbiscan® (5-7).

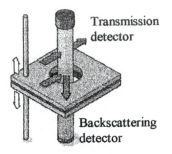

Transmission
detector

Backscattering
detector

Figure 5. Principle of Turbiscan® measurement   *Figure 6. Turbiscan LAb*

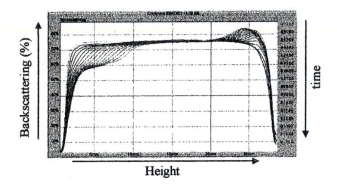

Figure 7. Superposition of scans in time for an unstable sample

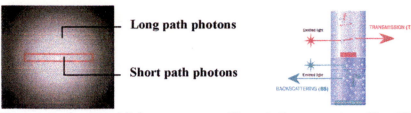

Long path photons

Short path photons

Figure1. Backscattered light spot

Figure 2. Representation of l and l*

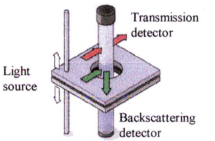

Transmission
detector

Light
source

Backscattering
detector

Figure 5. Principle of Turbiscan® measurement

Figure 6. Turbiscan LAb

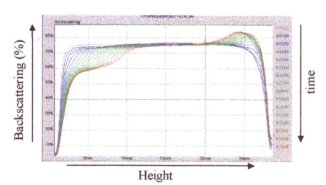

Backscattering (%)

time

Height

Figure 7. Superposition of scans in time for an unstable sample

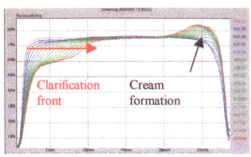

Figure 8. Creaming emulsion

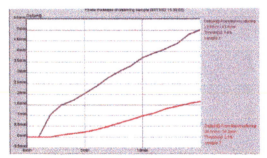

Figure 9. Comparison between samples

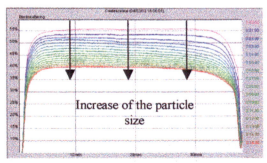

Figure 10. Coalescing emulsion

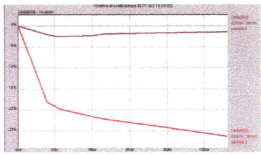

Figure 11. Comparison of samples

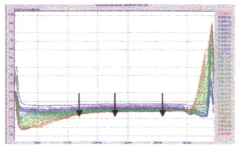

Figure 12. Emulsion flocculating and creaming (profiles in reference)

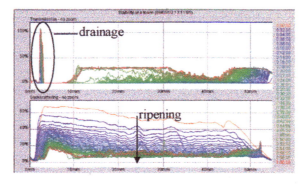

Figure 14. Breaking of a "metastable" foam (scanning mode)

Particle Migration - Creaming

Creaming is a common instability phenomenon encountered for emulsions when the dispersed phase has a lower density than the continuous phase. It can be coupled to coalescence or flocculation and will often lead to a phase separation. Creaming is detected as a result of concentration change between the top and the bottom of the measurement cell (Figure 8).

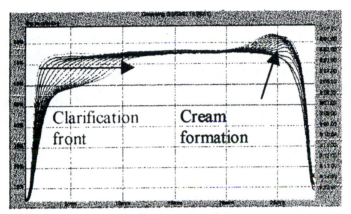

Figure 8. Creaming emulsion

We may clearly observe a decrease of the backscattered light flux at the bottom of the sample due to lower particle volume fraction in this region (clarification) and, on the other hand, an increase of the backscattering level at the top of the sample because of particle density increase during creaming. It is then possible to analyze creaming kinetics through the time evolution of the cream thickness and then compare the stability between samples.

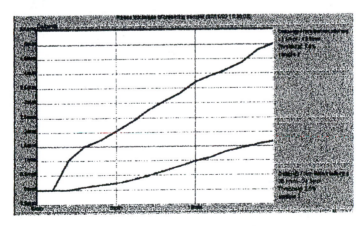

Figure 9. Comparison between samples

Figure 9 indeed shows that sample 2 (top) is more stable than sample 1 (bottom). The slope of the curve can be derived to give information about the migration rate.

The same kind of behavior is observed for sedimentation processes with an increase of the backscattering level at the bottom and decrease at the top of the sample.

Particle size variation

Coalescence and flocculation phenomena are physico-chemically very different but lead to an increase of the scatterers size. Both phenomena can be differentiated in the way that coalescence results from the fusion of closed drops whereas particles stick during a flocculation or aggregation process. In some case flocculation of droplets can lead to coalescence. Particle size variation induced by these phenomena is detected as it leads to a decrease of the backscattering level over the whole sample height (Figure 10).

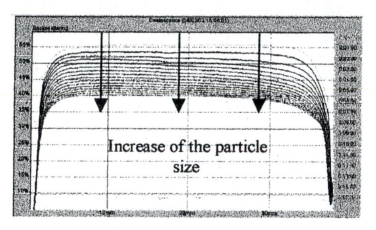

Figure 10. Coalescing emulsion

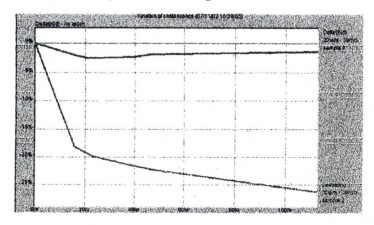

Figure 11. Comparison of samples

Figure 10 indeed shows a decrease of the backscattering level everywhere in the cell during a coalescence process (in the case of particles where d>0.7λ, Figure 3). The coalescence behavior of samples can be analyzed by plotting time evolution of the backscattering level (Figure 11) to enable comparison between formulations.

In industry, products do not generally undergo only one instability phenomenon but several at the same time. The Turbiscan® allowing a macroscopic visualization of the stability of concentrated dispersions, it is possible to discriminate various destabilizations. The profiles obtained, in this case, are a combination of the one previously described.

Depletion flocculation

One of the common instabilities taking place when formulating an emulsion is flocculation by depletion (*8*). Depletion forces can occur when colloidal species such as micelles, non-absorbing polymers or micro-emulsion coexist with larger droplets, since the small particles induce an osmotic pressure on the larger ones. If two emulsion droplets are brought together at such a distance that prevents the smaller species going in between the two, an uneven pressure is created. Eventually compression of the two emulsion droplets will occur, hence a flocculation phenomenon. One of the main characteristics of this instability is that it is reversible upon dilution as the depletion forces are low energy attractive forces. However this phenomenon can be quite difficult to observe, as a phase separation does not always appear and techniques such as particle sizer and microscopy disrupt the process.

The following experiment was carried out by homogenizing a hexadecane in water emulsion stabilized by sodium dodecyl sulfate (SDS) and running the characterization with the optical analyzer. As shown on the set of scans, Figure 12, the sample is unstable regarding flocculation (variation of the backscattering in the middle of the sample), leading to creaming. However, when diluted ten times in water, the sample becomes stable.

As the emulsion becomes stable when diluted it proves that it undergoes a depletion flocculation due to the high level of surfactant in the formulation.

Stability of foam

Foams consist of a dispersion of air bubbles in an aqueous phase. They can be present in the food industry, cosmetics, cements, surfactant chemistry, *etc.* However, they are quite difficult to characterise and few techniques exist to analyse them because of their low stability. The optical scanning analyser, Turbiscan®, can be used to study foams of various stability profiles:

    × Unstable foams, *i.e.* foams with lifetimes of seconds, can be studied *via* a fixed head option, where the emission-detection head is set at a fixed height and the evolution of the transmission and backscattering fluxes are followed with time (up to one acquisition every 0.1s). The breaking of the

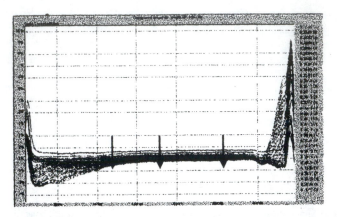

Figure 12. Emulsion flocculating and creaming (profiles in reference)

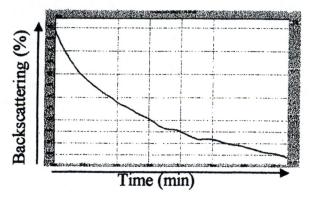

Figure 13. Breaking of an unstable foam (fixed mode)

air bubbles of the foam is characterised by a decrease of the backscattering (Figure 13).

× Metastable foams, *i.e.* foams that are stable for at least a few minutes before they break, can be studied through the "conventional" way, *i.e.* scanning of the sample. In this case, the backscattering profiles enable to measure the ripening of the air bubbles (bottom of Figure 14). Moreover, it is interesting to look at the transmission profile (top of Figure 14) and to follow the drainage of the foam (transmission peak at the bottom).

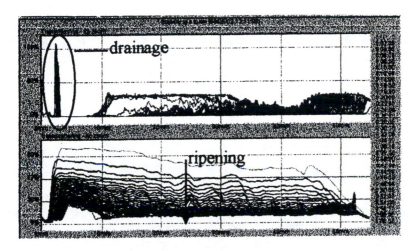

Figure 14. Breaking of a "metastable" foam (scanning mode)

When studying surfactants it is possible to use the apparatus to measure the foamability of a given surfactant and the stability of the foam formed through the same process. Therefore the optical scanning analyser, Turbiscan®, can be used to study the stability of foams and help the formulator to optimise formulations depending on the end-use properties (speed of breaking, foamability of a surfactant).

Conclusion

The optical analyser, Turbiscan®, is dedicated to stability analyses of liquid dispersions, diluted or concentrated, such as emulsions (simple or double), suspensions or foams. It is used in many different domains (food, cosmetics, inks, paints, petroleum, *etc.*). It can monitor and discriminate various types of instability phenomena, even if they take place simultaneously and before they are visible to the eye. It is also possible to accelerate the destabilisation by increasing the temperature *via* a thermo-regulated system.

Particle size measurement

Most particle size measurement techniques require a dilution in order to have single scattering. However, as seen previously, diluting a concentrated dispersion for characterisation can lead to a denaturation of the sample, hence the possibility to obtain results different from what is actually happening in the product.

In addition to the detection of instability phenomena, the optical analyser Turbiscan®, uses the MLS theory and the theories of Mie, Rayleigh and general optics, described in the first section of this chapter, to characterise dispersions in their native state. The analysis does not require any sample preparation (no dilution).

Dispersion state

The use of MLS enables to obtain information on the dispersion state of the product. It gives the parameter l^*, transport photon mean free path, which is directly related to the ratio d/ϕ (equation 3) and is therefore a fingerprint of the dispersion. This value only depends on the particle mean diameter, the volume fraction and the way particles interact together. It can be used as a quality control parameter to compare, for example, different batches of the same product or for scale up applications. Moreover, the vertical scanning of the sample gives the spatial distribution of l^*, hence the spatial distribution of d/ϕ of the product. This is an important parameter, as it represents the real dispersion state of the sample analysed. It will therefore give information on the homogeneity of the system.

Characterisation of a concentrated emulsion

If we consider a series of emulsions, consisting of mineral or silicon oil, at various volume fraction, in water and stabilised by a non-ionic surfactant, and we carry out a size measurement with particle sizers (D(4,3)) of two different brands and the Turbiscan® (mean diameter), we get the following results (Table 1).

Emulsion type	$d_{\text{Particle Sizer I}}$ (μm)	$d_{\text{Particle Sizer II}}$ (μm)	$d_{\text{Turbiscan}}^{®}$ (μm)
Hexadecane (20%)	1.45	1.52	1.72
Mineral oil (10%)	2.47	2.10	2.35
Silicon oil (30%)	6.43	6.35	6.27
Dilution	Yes	Yes	No

Table 1. Comparison of particle sizes of monodisperse emulsions

Characterisation of a flocculated emulsion

We have just seen that the Turbiscan® gives a mean diameter similar to what is obtained with classical particle sizer. However, these equipments require a dilution for the analysis that can have an effect on the sample analysed. Indeed, if we now consider an emulsion flocculated due to a depletion phenomenon (see section above), we know that the dilution performed during the particle sizer analysis will break the flocs. Hence, the emulsion will appear fine although it is flocculated. By doing the analysis with the optical analyser, we can show a flocculation phenomenon (Table 2).

In this experiment we have taken an emulsion, non-flocculated (checked by microscopy) and added some surfactant (sodium dodecyl sulfate) in excess. The surfactant forms micelles that provoke the depletion flocculation. By diluting the emulsion ten times the product is de-flocculated and we measure the initial diameter using the Turbiscan®, while the particle sizer does not see any difference between all samples due to the dilution that breaks the flocs.

Emulsion type	$d_{Particle\ Sizer}$ (μm)	$d_{Turbiscan}^{®}$ (μm)
Non flocculated emulsion	1.45	1.14
Flocculated emulsion	1.43	2.95
Diluted emulsion (x10)	1.39	1.32

Table 2. Comparison of particle size for a flocculated emulsion

Therefore, we can see the interest of using both techniques in the lab in order to fully characterise the dispersions as both of them have complementary assets: the particle sizer gives a particle size distribution and the optical scanning analyser is able to characterise the system in its real state, without any dilution.

Characterisation of a foam

As mentioned earlier in this chapter, it is very difficult to characterise foam stability, but it is even more complicated to measure bubble sizes in foams. Only microscopic techniques are used but they are time consuming (need of a large number of measurements to have statistically reliable data) and the sampling between lamellae can be particularly damaging for the products. Therefore, the use of the optical analyser, Turbiscan®, can be of great interest for the formulator or the analyst who wants to characterise foam.

Thanks to the sampling by coring, the product is not damaged and remains in its native state. It is also possible to form the foam directly in the cell. After the scan is done, we can get to the diameter of the bubbles by giving both refractive indices and the volume fraction.

Conclusion

The optical analyser, Turbiscan®, is therefore used as a particle size analyser as it gives the mean diameter of dispersions in their native state, without performing any dilution nor denaturation. Therefore, it can be used for quality control and analysis purposes to characterise colloidal products.

References

1. Kerker M. *The scattering of light*, **1969**, Academic Press, New York.
2. Snabre P., Arhaliass A., *Applied Optics* **1998,** 37 (18), 4017-4026.
3. Akira Ishimaru, Yasuo Kuga.. *J. Opt. Soc. Am.*, **1982,** 72 (10).
4. Gandjbackche A., Mills P., Snabre P. *Applied Optics,* **1995,** 35, 234.
5. Meunier G. *Spectra Analyse*, 1994, 179, 53.
6. Mengual O, Meunier G., Cayré I., Puech K., Snabre P. *Colloid and Surfaces A,* **1999,** 152, 111.
7. Mengual O, Meunier G., Cayré I., Puech K., Snabre P. *Talanta*, **1999**, 50, 445.
8. Aronson M.P. *Langmuir,* **1989,** 5, 49.

Chapter 4

Droplet Size Determination by Laser Light Scattering

Henrik G. Krarup

Malvern Instruments, 10 Southville Road, Southborough, MA 01772

Product delivery by spray has become increasingly important to the pharmaceutical, cosmetic, paint, fuel injector and agricultural industries. Droplet size is a key parameter in spray performance evaluation and nozzle design optimization. The current paper will illustrate how laser light scattering is used to characterize droplet size distributions in sprays. This is not a trivial matter since sprays often are highly concentrated, occupy large areas and may be pulsed at short frequencies. Thus the laser scattering instrument has to be able to cope with multiple scattering and provide rapid data acquisition. The method provides size resolution in the range from 0.5 – 850 micrometer at concentration levels up to 95 % light obscuration. Data may be acquired up to 2500 Hz.

Introduction

Laser Light Scattering or Laser Diffraction, as it is also commonly referred to, is a spatial technique. The size information is acquired instantaneously as a multitude of particles passes through the measurement zone and is displayed as a distribution of sizes as a function of occurrence in individual size bins (Fig. 1). The size distribution is characterized by percentiles as the median Dv50 and averages such as the Sauter Mean Diameter or the D[3,2]. The measurement is volume based as the diameter of a light scattering object is derived from the volume of a sphere with a volume equivalent to that of the object. Unlike high-speed photography, laser light scattering is a non-imaging technique. Yet other techniques derive size distributions as individual droplets are analyzed as they pass through a measurement zone during an interval of time. These techniques are referred to as flux techniques and examples hereof would be Number Counters and Phase Doppler Analyzers. Typically Number Counters are preferred in low concentration applications such as monitoring of airborne particles in clean rooms and hospitals. Laser Light Scattering on the other typically receives information from hundreds of thousands of particles during a measurement and applies equally well to continuous as transient sprays. Hence the technique covers applications such as pharmaceutical nebulizers [1,2], metered dose inhahalation sprays, dry powder inhalers, industrial fog [3], fuel injection sprays [4], agricultural electrostatic sprays [5], paint sprays, domestic and cosmetic sprays. For slow data acquisition at 1 Hz, results are shown real time. During faster acquisition up to 2500 Hz the results follows about 30 seconds after the measurement.

Instrument Design

The Laser Light Scattering Instrument requires a source of monochromatic, collimated light. As the light interacts with matter, scattering phenomena occurs and the scattered light is focused onto the detectors via a Fourier Transform Lens (Fig.1). The lens focuses scattered light from droplets of similar size to the same detector position irrespective of their velocity and position in the measurement zone. The raw data obtained from such an instrument is thus the intensity of the scattered light as a function of the detector position.

From raw data to results

The general theory of light scattering by isotropic spherical particles of any size as formulated by Gustav Mie [6] in 1908 enable us to calculate the scattering amplitude function from a given size distribution, with specified refractive indices and known wavelength of the experimental light source. To

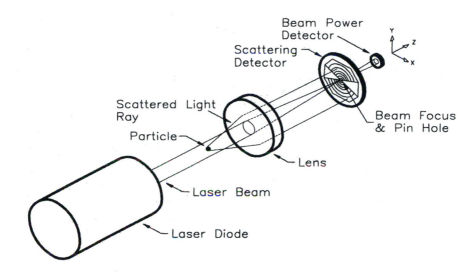

Figure 1. Schematic of a laser light scattering instrument. The laser has a wavelength of 670 nm and a beam diameter of 10 mm. 3 different range lenses provide information in the size range from 0.5 – 850 μm. Each lens has a working distance which is a function of the focal length. In order to avoid vignetting i.e. loss of scatter signal from the smallest droplets within the distribution, all droplets must be within the specified working distance from the lens. Purge air may be provided to avoid deposition of spray material on the optic surfaces.

convert the observed scattering amplitude function to a droplet size distribution the instrument utilizes the fact that droplets scatter light at angles inversely proportional to their sizes. This allows the instrument to identify an initial guessed particle size distribution from which a calculated scattering amplitude function is generated. An iterative algorithm is developed in which the initial particle size distribution is manipulated until minimization of the difference between the observed and the calculated scattering amplitude functions is obtained (Fig.2).

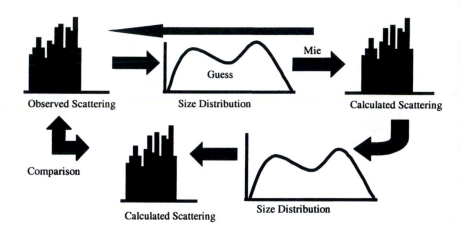

Figure 2. Schematic of algorithm used to derive the particle size distribution from the observed scattering amplitude function.

Multiple Scattering

In an ideal situation, the light scattered by any particle in a cloud of particles will pass directly from that particle to the detector. However as the concentration of particles increases, a phenomenon known as "multiple scattering" becomes apparent and this poses a special problem. Multiple scattering occurs when the light scattered from a particle is re-scattered by one or more subsequent particles on its way to the detector. The result of this effect is that light tends to be scattered to higher angles than theory predicts - resulting in an underestimation of the particle size. This has always been a particular problem in the sizing of concentrated sprays where multiple scattering is always present. Unfortunately, multiple scattering is an insidious process and it is sometimes difficult to discover at which concentration it becomes problematic for a given ensemble of particles. Van de Hulst (7) has stated that multiple scattering occurs when particles are closer than three times the radius of the particle. Hence it is entirely dependent on the size of the particles. Small particles will exhibit multiple scattering at significantly lower concentrations than larger particles. Virtually any spray will have a degree of multiple scattering present at light transmissions of 80% (obscurations of 20%). Sprays of small droplets will exhibit multiple scattering at much lower concentrations.

The problem of multiple scattering has been solved by the development of a correction algorithm that can accurately measure particle size distributions with light transmissions as low as 5% (obscurations of 95%):

$$A_{ctual} = P_1 S_1 + P_2 S_2 + P_3 S_3 + P_4 S_4 + P_5 S_5 + P_n S_n + \ldots + P_\infty S_\infty$$

The effect can be viewed as a convolution of the probability of n scattering events with the scattering signature for n events. This patented probability-based approach was designed by Harvill and Holve[8]. By solving the above equation and substituting the result into the normal scattering equations, the particle size distribution can be corrected for multiple scattering.

Beam Steering

Some sprays such as pharmaceutical pressurized metered dose inhalation (pMDI) sprays contain drive gas to deliver the active ingredient. As the drive gas evaporates the associated refractive index gradients deflects portions of the scattered light towards the inner rings of the detector system (Fig.3). This leads to the interpretation of phantom large particles (Fig. 4).
The phenomenon of beam steering [9] is remedied by killing the signal on the inner rings (Fig. 5,6).

Measurement accuracy

The droplet size depends on the flow rate/pressure on the nozzle and as such measurement repeatability varies more for sprayed droplets than for solid particles. Figure 7 illustrates the error function associated with measurement on a reference reticle used to verify the performance of the instrument. The reticle consists of a glass slide with a known population of particles deposited on top.

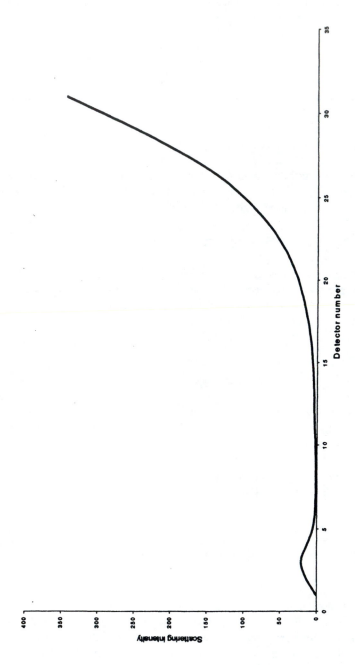

Figure 3. pMDI spray detector response – notice small peak between detectors 3-6. This population causes the occurence of the large particles observed on Fig.4.

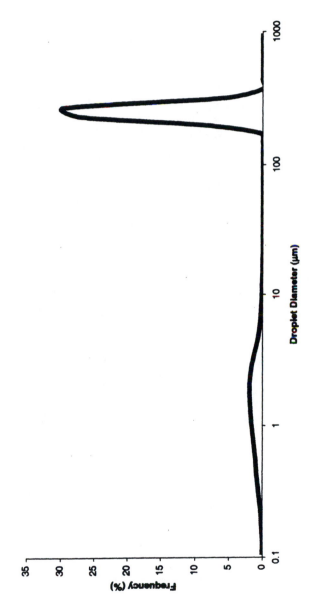

Figure 4. pMDI spray - Phantom large droplets caused by beam steering.

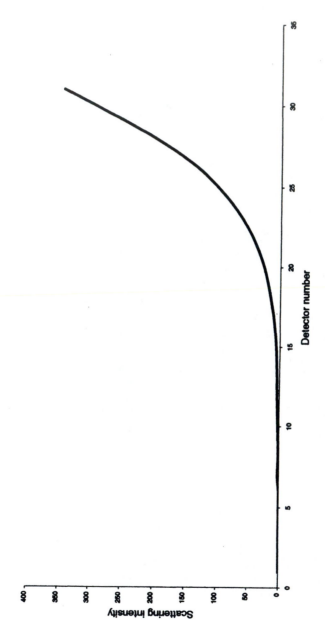

Figure 5. pMDI spray: Compare to Fig. 3 - small peak on detectors 3-6 has been removed.

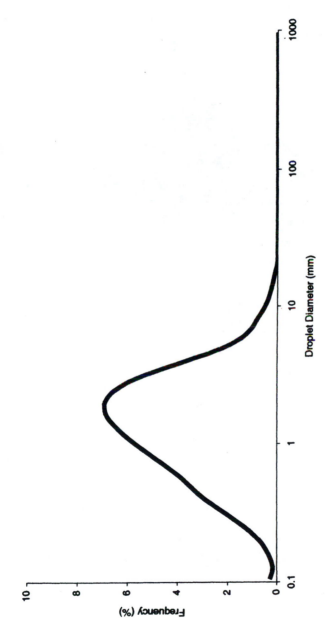

Figure 6. pMDI spray: Resulting droplet size distribution after data on the first 6 detector channels have been killed.

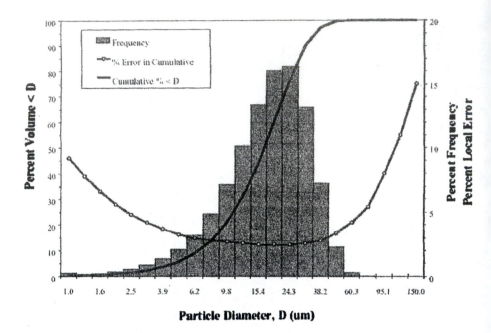

Figure 7. Error function for the Laser Light Scattering instrument associated with measurement on a static reference reticle. In the region between Dv10 and Dv90 the error on the measurement is on the order of a couple of percent.

Measurement Examples

The time history display (Fig. 8) combines information on the dynamics of the spray event monitored by the transmission curve and the droplet size distribution at a given time identified by 3 size parameters. Data was acquired at 1000 Hz and thus the data displayed represents 140 individual detector readings/droplet size distributions. Averages (Fig. 9) may be performed on any amount of time segments

Other spray applications, such as fuel injectors, pulses at high frequencies and thus demand high data acquisition rates from the Laser Light Scattering instrument (Fig.10).

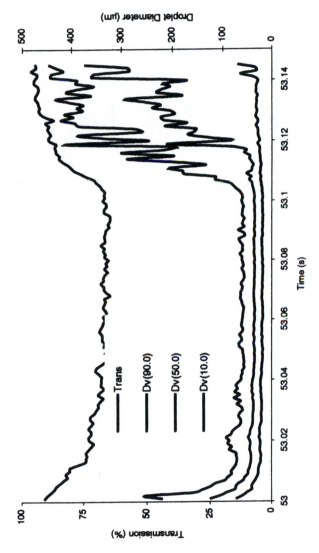

Figure 8. Time history presentation of nasal spray event lasting 140 ms.

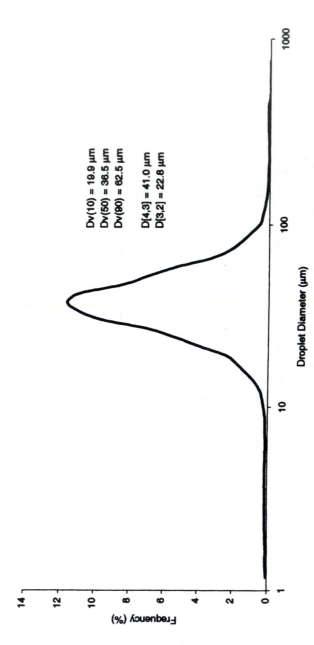

Figure 9. Droplet size distribution based on an average of the 100 ms, fully developed nasal spray event displayed above.

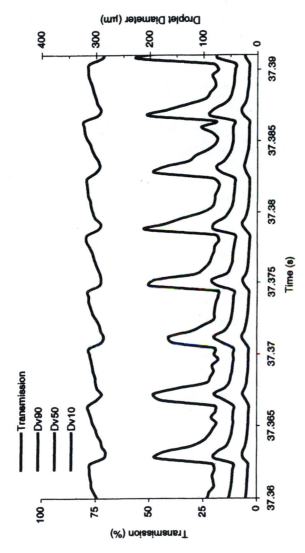

Figure 10. Time history data display of a pulsing fuel injector spray. Data was acquired at 2500Hz. Each pulse is identified by a drop in the upper transmission curve and corresponding peaks of the size parameters curves. The width of the pulses is 2 ms and each pulse is thus defined by 5 data points.

Summary

Laser Light Scattering has evolved into a position of offering fast and robust measurements of droplet sizes to the many and varied spray applications. Data may be displayed real time or after acquisition in time history displays or as droplet size distributions. Data may be acquired at 2500 Hz at 95% light obscuration that accommodate the fast pulsing sprays such as fuel injectors.

References

1) Kwong, W. T.; Ho, S. L. and Coates, A. L.; J. Aerosol Med., 2000, 13(4), p. 303-14.

2) Corcoran, T.E.; Hitron, R.; Humphrey, W. and Chigier, N., J. Aerosol Sci. 2000, Vol. 31, No. 1, pp. 35-50.

3) Chaker, M; Meher-Homji, C. B. and Mee III, T.; Proceedings of ASME Turbo Expo 2002, June 3-6, 2002 Amsterdam, 2002-GT-30562, 2002-GT-30563, 2002-GT-30564.

4) Robart, D.; Breuer, S.; Reckers, W. and Kneer, R. Assessment of Pulsed Gasoline Fuel Sprays by Means of Qualitative and Quantitative Laser-based Diagnostic Methods.

5) Jahannama, M. R.; Watkins, A. P. and Yule, A. J.; Proceedings of ILASS-Europe'99, p. 1-6, Toulouse 5-7 July, 1999.

6) Mie, G. Annalen der Physik, Vierte Folge, Band 25, 1908, No. 3, S. 377-445.

7) Hulst, H.C. van der: "Light Scattering by small particles", Dover Publications Inc., New York, 1981, 470 p.

8) Harvill, T.L. and Holve, D.J. (1998):"Size Distribution Measurements under Conditions of Multiple Scattering with Application to Sprays", Proceedings of ILASS 98, 5 p.,Sacramento,CA

9) Williams P.A. (1993): "Laser Diffraction Particle Sizing and Spark-Flash Photography Techniques for Successful Measurement of Spark-Ignition Engine Fuel Sprays". Proc. IMechE Conf. on Experimental and Predictive Methods in Engine Research and Development, 17-18th November 1993, NEC Birmingham, UK, pp. 261-272. IMechE Publication No. C465/016/93. ISBN 0 85298 862 1.

Chapter 5

Droplet Size Distribution of Oil–Water Emulsions by Confocal Laser Scanning Microscopy

José M. Benito, Guillermo Ríos, Carmen Pazos, and José Coca[*]

Department of Chemical and Environmental Engineering, University of Oviedo, 33006 Oviedo, Spain

A brief description of some methods used for droplet size analysis is presented. Special attention has been paid to the determination of droplet size distribution (DSD) of oil-in-water emulsions, used as water-based coolants, and destabilized by addition of $CaCl_2$. The emulsions were gelled in agarose so that the oil droplets were immobilized and samples of these gels were later characterized with a system of confocal laser scanning microscopy (CLSM) and image processing. The oil droplets were immobilized within the gel and no distortion of the droplets was observed because of the gelling process. The use of the CLSM technique, along with image processing analysis, allows one to follow the coalescence of oil droplets during the destabilization process and the remaining secondary O/W emulsion, which causes a residual turbidity.

Oil in water (O/W) emulsions appear in many industrial processes (food, paint, and cosmetic manufacture) and are generated in oil refining and petrochemical processes. They are also used for cooling and lubrication purposes in metalworking operations. Large amounts of O/W emulsions are used in the steel and aluminum industries in hot and cold rolling.

Treatment and recycling of O/W emulsions is a complex process which may involve coagulation, flocculation, gravity separation, fixed bed adsorption and membrane techniques (*1*). The selection of the treatment process depends on the droplet size distribution as one of the key parameters. Droplet size distribution also affects the rheological properties of the emulsion and its stability and resistance to creaming.

In this work, a brief description of methods for droplet size analysis and distribution of O/W emulsions by confocal laser scanning microscopy (CLSM) and image processing will be given. The applications and measurements using CLSM to characterize oil/water emulsions of the type used as cutting oils in mechanical operations will be presented.

Methods for Droplet Size Analysis

Techniques for measuring the droplet size distribution (DSD) in emulsions may require batch sampling and dilution of the sample, processes that usually alter the true distribution, or *in situ* measurements which are always preferable because they eliminate the need to collect samples and with subsequent external analysis. Another option is to use on-line sampling techniques, with the use of glycerine, which thickens or gels the emulsion (*2*), in order to reduce the effects of Brownian motion, coalescence and creaming of the oil droplets.

There are a vast number of methods for determining the DSD in emulsion systems. Overviews with details of these methods are available in the literature (*2-6*) and only a brief description of some of the most commonly used will be presented in this section. These methods can be broadly divided into four categories (*3*) based on the following observations: differences in electrical properties between the dispersed (oil) and the continuous (water) phases, scattering phenomena due to the presence of the dispersed phase, physical separation of the dispersed droplets, and microscopy.

Sizing by Electrical Properties

A widely used technique based on differences in electrical properties is performed with equipment often referred to as *Coulter counter* because the Coulter Corporation (formerly Beckman Coulter) manufactured the first commercial apparatus. It uses the change in electrical resistance when a single droplet (or particle) diluted in an electrolyte passes through two electrodes (sensing zone), being this change proportional to the amount of electrolyte displacement and therefore to the volume and size of the droplet. It is a well established technique for the size range from 0.6 to 1200 μm (4). This technique and its variations that allow to measure current, capacitance, or voltage changes, are also known as a *sensing zone techniques*. Such techniques always require calibration and dilution of the emulsion, and they are limited to oil-in-water emulsions with no solids present (3).

Sizing by Light Scattering

Most of the commercial sizing instruments are based on measuring the intensity of the scattered light, which depends among other factors on the size and shape of the droplets (or particles) in the emulsion. The advantage of light scattering techniques is that droplet size measurements can be made in any solvent but, furthermore, dispersions must be dilute. Laser light is preferred, as it is monochromatic, stable and intense. These techniques can be broadly divided into static and dynamic light scattering (7).

Static light scattering is used to determine droplet sizes in the range of 0.1 to 1000 μm (5). Measurement techniques based on static light scattering include:

- *Angular scattering methods,* which use a He-Ne laser to measure the angular dependence of the light scattered by the emulsion droplets (3, 8).
- *Spectroturbidimetry,* which measures the turbidity of a dilute emulsion as a function of wavelength, by comparing the intensity of light which has passed directly through the emulsion droplets with that which had passed directly through the continuous phase (9, 10).
- *Reflectance methods.* This technique has been used to determine the DSD of concentrated emulsions by measuring the light reflected back for the emulsion droplets (8, 11).

Dynamic light scattering allows determination of droplet sizes between 3 nm and 3 μm, below the lower limit of detection of static light-scattering techniques (12). Methods based on dynamic light scattering include Fraunhofer diffraction and light scattering at larger angles (Mie scattering). Figure 1 shows a schematic view of a light scattering apparatus with examples of the observed signal: turbidity can indicate the amount of dispersed phase, but offers no information about droplet size distribution (3); Fraunhofer diffraction offers information on size when particles or droplets of the dispersed phase are of the order of, or larger than, the wavelength of the incident light (3, 4). Mie scattering occurs when particles or droplets are smaller than the wavelength of the incident light (4, 13).

The most commonly used method in commercial instruments, based on dynamic light scattering, is *photon correlation spectroscopy* (PCS), also known as *quasi-elastic light scattering* (QELS). PCS is based on the scattering pattern produced as a result of the interaction between the electromagnetic waves of the laser light and the oil droplets (or particles of the dispersed phase). Scattered-light intensity fluctuations with time, as a result of the Brownian motion of droplets, depends on the speed at which the droplets move, and hence on their size. PCS implies the measurement of the scattered light intensity at a given angle (*e.g.*, 90°) by a photomultiplier detector in a series of short time intervals and their digital correlation (*7, 14, 15*). This technique is restricted to the analysis of dilute emulsions with droplet sizes between 0.005 and 3 μm.

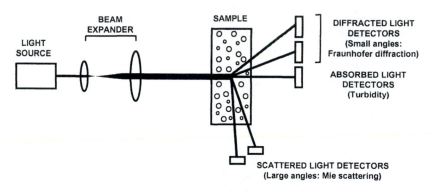

Figure 1. Set-up of a light-scattering apparatus. (Adapted from reference 3. Copyright 1992 American Chemical Society.)

Sizing by Physical Separation

Mechanical separation/classification techniques (*i.e.*, sieving, filtration, elutriation, etc.) generally involve systems of solid particles, but some developments in physical separation techniques, such as chromatography, sedimentation and field-flow fractionation have applications to emulsion systems.

• *Chromatographic techniques.* Hydrodynamic and size-exclusion chromatography are two techniques employed for the separation and size measurement of emulsions and micellar systems. Both techniques are based on conventional chromatographic principles either of flow of particles or droplets in a carrier fluid. The larger droplets are eluted from the chromatographic column first because they have not interactions with pores in the packing material (size-exclusion chromatography), or because they are too big to stay in the lowest velocity zones, near the non-porous packing or at the walls of the column (hydrodynamic chromatography). Typical droplet sizes range up to 1 µm (*2, 3*).

- *Sedimentation.* Gravitational and centrifugal sedimentation, in which particles or emulsion droplets are separated on the ability of gravity force to distinguish among different sizes, can be used to determine size distributions of droplets between 1 nm and 1 mm. The rate at which droplets sediment, or cream, in a gravitational or centrifugal field may be monitored using X-ray or light absorbance as a function of position, optical microscopy, nuclear magnetic resonance, ultrasound, and electrical measurements (*3, 7, 16*). The DSD may be determined by analyzing the droplet velocity, using the Stokes equation for gravitational and the sedimentation coefficient for centrifugal sedimentation, respectively (*5*).

- *Field-flow fractionation.* This technique combines hydrodynamic chromatography and a gravitational, electric, magnetic, thermal, centrifugal, or whatever force field that might interact with the droplets or particles (*2, 3, 7*). Droplets are sized according to their response to the field, which is applied at right angles with respect to the emulsion flow. Sedimentation and thermal field-flow fractionation are the most commercially available techniques. Field-flow fractionation techniques have been used to determine particle diameters ranging from 0.001 to 100 µm (*4*).

Microscopy

The most common method for droplet sizing is observation and counting of the droplets by microscopic techniques. Microscopy can be regarded as the single most important emulsion characterization tool (*3*). Various microscopic techniques are available for analyzing the structure, physical nature and composition of emulsions: optical microscopy, electron microscopy, and confocal laser scanning microscopy. Traditionally, optical microscopy has been a widely used technique, but because of the errors associated with it (*2*), image processing through a computer interface yields far more reliable results. Nonetheless, there are problems associated with image processing, because there is a high probability that closely spaced aggregates can be counted as a single unit, and large aggregates may mask the presence of smaller droplets that are either positioned behind or superimposed on the large aggregate. These problems can be easily minimized if the dispersed phase concentration is low

(*17*). Methods exist for separating individual droplet images, such as the inscribed circle method (*18*).

Some parameters have to be carefully considered when a commercial microscopic instrument is used for the characterization of an emulsion, namely resolution and contrast (*2, 5*). *Resolution* is the capability to distinguish small separations between two objects, and is defined by the wavelength of the illumination, the refractive index of the medium, and the angle of light acceptance by the lens. *Contrast* determines how well an object can be distinguished from their surrounding medium, and it depends on the relative refractive index of the two substances.

Conventional Optical Microscopy

Optical microscopy if one of the most valuable tools for observing the microstructure of emulsions. It involves the use of transmitted, reflected, and polarized light, fluorescence, and more recently, techniques such as confocal microscopy (*3, 5*). Each of these variations has particular strengths and applications.

An optical microscope contains a series of lenses that direct the light through the emulsion and magnify the resulting image. The theoretical limit of resolution of an optical microscope is about 0.2 μm, but in practice it is difficult to obtain reliable measurements below about 1 μm, due to the Brownian motion of small droplets (*4, 5, 8*). The characteristics of particles with sizes less than one micrometer can be observed by dark-field illumination using an ultramicroscope (*4, 8*).

Transmitted-light microscopy requires a sample sufficiently thin to allow light to pass through it, and it is not applicable for optically opaque samples. In cases where the sample cannot be made thin enough, an alternative technique, reflected-light microscopy, is available (*3*).

Fluorescence and Confocal Laser Scanning Microscopy (CLSM)

Fluorescence microscopy is mainly applicable to oil-in-water emulsions in which the oil phase appears as bright spots in a dark background (water does not fluoresce). This technique is particularly suitable for image analysis and automated droplet counting and size characterization (*3, 19*).

Based on fluorescence microscopy, confocal laser scanning microscopy (CLSM), also known as LSCM, offers several advantages over conventional optical microscopy, including higher clarity images, as out-of-focus interferences were eliminated, and the ability to collect serial optical sections for the generation of three-dimensional structures without the need of physically section the sample (*5*). The invention of the confocal microscope is attributed to Marvin Minsky, who built a working microscope in 1955 and patented the principle of confocal imaging in 1957.

The basic principle of CLSM is shown in Figure 2 (*20*). A point in the sample being analyzed is illuminated with one or more focused laser beams. The point of light is reflected by a dichroic mirror and is focused by an objective lens at the desired focal plane in the sample. Fluorescence is excited in the sample and emitted in all directions. Only the fluorescence that originates in the plane of focus (in-focus light) reaches a second pinhole having the same focus as the first pinhole (the two are confocal). Any light that passes the second pinhole reaches a photomultiplier, which generates a signal that is related to the brightness of the light from the sample. The second pinhole prevents light originating from above or below the plane of focus in the sample from reaching the photomultiplier (*20*). A two-dimensional image is obtained by carrying out measurements at the *x-y* plane and by combining the measurements from each individual point. Three-dimensional images are obtained by focusing the laser beam at different depths into the sample (*z*-series) and then scanning in the horizontal plane. The key feature of the CLSM is that only what is in focus is detected, due to the use of spatial filtering to eliminate out-of-focus light or flare. The out-of-focus parts appear as a black background in the final image.

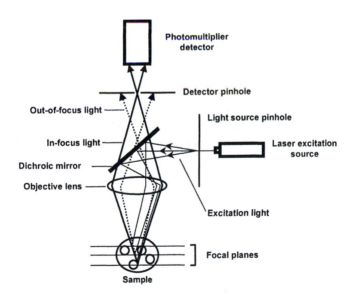

Figure 2. Diagram of the confocal laser scanning microscopy principle.

Electron Microscopy

Scanning (SEM) and transmission (TEM) electron microscopy are widely used to characterize emulsion systems (*21, 22*), specially those that contain structural components which are smaller than 1 μm (*5*). The practical lower limit of emulsion sizing with electron microscopy is on the order of 1 nm (*3, 5*). Electron microscopes use electron beams, rather than light beams, to provide information about the structure of emulsions, and use a series of magnetic fields rather than optical lenses. The major disadvantages are the effect of high vacuum and the high temperature created by the electron beam on the sample, which alter its state either by flocculation or coalescence (*4*).

Other Techniques

Several methods for droplet size measurement have been developed in recent years, specially for their use with opaque and concentrated emulsions or dispersions. Some of them, such as nuclear magnetic resonance (NMR) spectroscopy, are applications of well-established technologies to emulsion systems. A NMR technique has been developed to measure the DSD of concentrated O/W and W/O emulsions, and it is sensitive to droplet sizes between 0.2 and 100 μm (*3, 5*).

Recent research has been carried out using acoustic-sizing techniques. The main advantage of these techniques is that they can be used with optically opaque and/or concentrated emulsions or dispersions *in situ*, even through metal walls (*i.e.*, pipes, tanks or reactors). Particle-sizing instruments based on ultrasonic spectrometry utilize interactions between ultrasonic waves and particles to determine droplet sizes between 10 nm and 1000 μm (*4, 23*). The two major drawbacks of ultrasonic spectrometry are the large amount of thermophysical data required to interpret the measurements and the fact that small air bubbles can interfere with the signal from the emulsion droplets (*24, 25*). Nevertheless, ultrasonic spectrometry has gained increasing acceptance in some industrial processes, specially in the food industry (*5*).

Droplet Size Distribution of O/W Emulsions by CLSM

This section deals with the use of CLSM and image processing analysis in connection with the study of the destabilization process of an O/W emulsion. Gelling of the emulsion facilitates its study once the emulsion has been destabilized (26). The oil droplets were immobilized within the gel and no distortion of the droplets was observed due to the gelling process.

A commercial miscible cutting oil Alba-Kool (Berkol Co. Spain) was used as provided by its manufacturer. The inorganic salt used to destabilize the emulsions was $CaCl_2$ (Panreac, reagent grade). The gelling agent was Agarose Type I-B: Low EEO (Sigma-Aldrich).

Fluorescence microscopy measurements were carried out using a CLSM Bio-Rad MRC-600, with an IMCO 10 image processor. MIP and SMUI software (Microm S.A.) were used for image processing and morphometry.

Sample Preparation and CLSM Measurements

3 vol % emulsions of the commercial oil in distilled water were prepared at room temperature and stirred at 250 rpm for 10 min using a magnetic stirrer. To determine the influence of $CaCl_2$ on the droplet size distribution, 2.5 g of salt were added to 250 mL of the emulsion, so that a salt concentration of 10 g/L was obtained.

As it was mentioned above, oil/water emulsions were gelled in agarose to immobilize the oil droplets. In order to prepare the gel, a solution of 2 wt % agarose in distilled water was placed in a microwave oven (the temperature necessary to dissolve the agarose is ca. 80 °C). Once the optimum agarose concentration to be used was determined, several trials were carried out to ascertain the required ratio of volume of agarose solution to volume of the emulsion sample, namely 1:1. The sequence in which the agarose solution and the emulsion sample are added influences the uniformity of the mixture. The best way is to first add the agarose solution to a test-tube and then add the emulsion sample.

The experimental protocol is as follows (27): 2 mL of hot agarose solution were placed in a plastic test-tube; 2 mL of emulsion sample were then added and the mixture was allowed to cool until the gelling temperature (35-40 °C) was reached and a gel was formed in the test-tube. The gel was removed from the tube by breaking its top, and it was then cut in circular portions using a scalpel. These circular portions were placed on a microscope slide and stored in a refrigerator to minimize evaporation of water. Finally, size distributions of the samples were measured using CLSM and image processing. The use of fluorescence microscopy makes the droplets appear as bright spots on a dark background and avoids the problems associated with the presence of air bubbles, which may have been trapped in the gel (19).

Image Processing Analysis

All captured images had similar characteristics and in order to improve the original image and to identify and characterize the droplets present in the sample, several processing steps are required:

1. *Image acquisition.* Images captured from a CLSM were converted into digital format. A digital image is composed of *picture elements* or *pixels.* Each pixel has a grey value (*i.e.* from 0 (black) to 255 (white), in a 8-bit greyscale confocal image) depending on the fluorescence intensity at a point within the sample. A single image, called *grey image*, was obtained from maximum values of each pixel for all of the images in the same series.

2. *Image enhancement.* In this step the grey image was processed to improve its resolution and quality. This processing step involves application of LUT (*look-up table*) transformation and median filter, and the result is an image with nearly uniform regions that are effectively classified for segmentation.

    -   LUT transformation. Contrast and grey values were altered by using mathematical operations, without modifying the original grey image.
    -   Median filter. This mathematical algorithm was used to eliminate noise and to correct for problems with brightness. Each pixel of the filtered image is defined as the median brightness value of its corresponding neighborhood in the original grey image.

3. *Segmentation.* The desired objects (oil droplets) were extracted from the background. Starting from a threshold grey value, a *binary image*, in which each object appears in white and the background in black, was obtained.

4. *Morphological operations.* The binary image was subjected to the following operations:

    -   Closing. Pixels that appeared in black in contact with a majority of white pixels were converted to white and *vice versa*. This operation allows one to soften the shape of the objects.
    -   Filling. Black holes within white objects were converted to white.
    -   Expert sieve. Objects which were not essentially circular in shape or not in contact with the frames of the image, were eliminated.

5. *Calibration.* Appropriate patterns are introduced so that the measurements in pixels can be converted to microns, millimeters, etc. In some cases, morphological operations allow to distinguish between oil droplets initially aggregated.

6. *Measurement.* Once every object has been identified, different morphological parameters (projected area, maximum diameter, different equivalent diameters, etc.) were measured or calculated and saved in a file.

Once the individual droplet sizes were obtained, the data were classified in order to obtain the droplet size distribution. Classification is accomplished by arranging the measured values into a number of classes, defined by two droplet diameters known as class boundaries. Because the size range is quite wide, a log-normal-type distribution is expected. Thus, data must be plotted on a functional scale (regular intervals with equal differences between the logarithms of the class boundaries; the geometric mean thus represents the average class diameter). In order to preserve the morphology of the distribution, it is recommended the divisions to be chosen so that the class interval divided by the average class diameter is approximately constant. Either 11 or 12 classes were selected because it is recommended to use between 10 to 20 classes (*28*).

The number of droplets in each class (n_i) and the corresponding percentage were determined by plotting the raw data as a frequency distribution. Because the droplets are spherical, the equivalent diameter was taken as the actual diameter, D_i (*29*). This distribution may be converted into other frequency distributions, such as the percentage of the surface area or volume corresponding to a specific diameter. For isomorphic particles, the percent by volume may be derived from the percent by number, $((n_i/\Sigma n_i)\times100)$, as $((n_iD_i^3/\Sigma n_iD_i^3)\times100)$.

Results and Discussion

The evolution of droplet size distribution with time can be qualitatively observed using grey images obtained by CLSM (Figure 3). Oil droplets up to 60 μm were observed 5 minutes after the CaCl$_2$ addition, this size being lower for higher destabilization times. Droplet size distributions for different O/W emulsions using CLSM and photon correlation spectroscopy (PCS) techniques were discussed in a previous work (*27*), obtaining similar results for the destabilized emulsions, with a good reproducibility.

Droplet size distributions obtained by CLSM and their variation along time are shown in Figures 4 and 5. When the inorganic salt (CaCl$_2$) was added, the electrostatic forces between oil droplets become negligible, and therefore the size distribution is shifted to larger sizes. Furthermore, due to the coalescence of larger oil droplets, size distribution was shifted again to smaller sizes with time, because only small droplets remain in the emulsion. Likewise, the decrease of the numeric concentration of oil droplets causes the formation of a secondary oil/water emulsion, which is responsible for the residual turbidity observed at the end of the destabilization process.

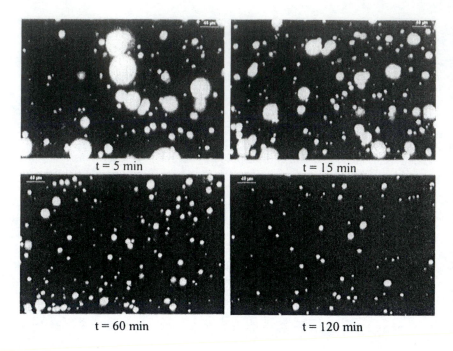

Figure 3. Grey images of the Alba-Kool O/W emulsion after CaCl₂ addition.

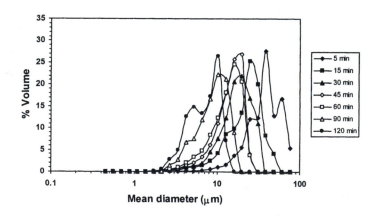

Figure 4. Variation of droplet size distribution with time for Alba-Kool emulsion, after the addition of 10 g/L CaCl₂ using CLSM.

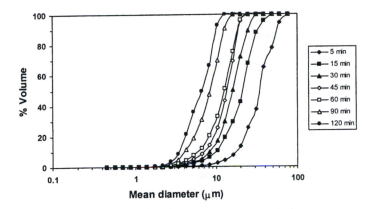

Figure 5. Variation of droplet size distribution with time (cumulative percent) for Alba-Kool emulsion, after the addition of 10 g/L CaCl₂ using CLSM.

Conclusions

A short review on the imaging techniques for the characterization of emulsions has been presented. Droplet size distributions for oil-in-water emulsions have been measured by immobilizing the oil droplets in agarose and using a method of confocal laser scanning microscopy (CLSM) and image processing analysis. The analyzer is capable of detecting droplets below 1 µm and it allows to follow the destabilization of the emulsion with CaCl₂ as coagulant.

References

1. Benito, J. M.; Ríos, G.; Pazos, C.; Coca, J. In *Trends in Chemical Engineering;* Research Trends: Trivandrum, India, 1998; Vol. 4, pp 203-231.
2. Orr, C. In *Encyclopedia of Emulsion Technology;* Becher, P., Ed.; Marcel Dekker: New York, 1988; Vol. 3, pp 137-169.
3. Mikula, R. J. In *Emulsions. Fundamentals and Application in the Petroleum Industry;* Schramm, L. L., Ed.; Advances in Chemistry Series 231; American Chemical Society: Washington, DC, 1992; pp 79-129.

88

4. Morrison, I. D.; Ross, S. *Colloidal Dispersions. Suspensions, Emulsions, and Foams;* Wiley-Interscience: New York, 2002; pp 62-116.
5. McClements, D. J. *Food Emulsions. Principles, Practice, and Techniques;* CRC Press: Boca Raton, FL, 1999.
6. Barth, H. G. *Modern Methods of Particle Size Analysis;* Wiley-Interscience: New York, 1984.
7. Hiemenz, P. C.; Rajagopalan, R. *Principles of Colloid and Surface Chemistry;* Marcel Dekker: New York, 1997.
8. Farinato, R. S.; Rowell, R. L. In *Encyclopedia of Emulsion Technology;* Becher, P., Ed.; Marcel Dekker: New York, 1988; Vol. 1, pp 439-479.
9. Walstra, P. *J. Colloid Interface Sci.* **1968**, *27*, 493-500.
10. Reddy, S. R.; Fogler, H. S. *J. Colloid Interface Sci.* **1981**, *79*, 101-104.
11. Lloyd, N. E. *J. Colloid Interface Sci.* **1959**, *14*, 441-451.
12. Dalgleish, D. G.; Hallet, F. R. *Food Res. Int.* **1995**, *28*, 181-193.
13. Hofer, M.; Schurz, J.; Glatter, O. *J. Colloid Interface Sci.* **1989**, *127*, 147-155.
14. Cummins, P. G.; Staples, E. J. *Langmuir* **1987**, *3*, 1109-1113.
15. Horne, D. S. *J. Colloid Interface Sci.* **1984**, *98*, 537-548.
16. Pal, R. *Colloids Surf. A* **1994**, *84*, 141-193.
17. Farrow, J.; Warren, L. In *Coagulation and Flocculation. Theory and Applications;* Dobiáš, B., Ed.; Marcel Dekker: New York, 1993; pp 391-426.
18. Koizumi, F.; Kunugita, E.; Nishitani, H. *Proc. PSE'94* **1994**, 747-752.
19. Sampedro, A.; de los Toyos, J. R.; Martínez-Nistal, A. *Técnicas de Fluorescencia en Microscopía y Citometría;* Universidad de Oviedo, Servicio de Publicaciones: Oviedo, Spain, 1995.
20. *Introduction to Confocal Microscopy,* URL http://micro.magnet.fsu.edu
21. Mikula, R. J.; Munoz, V. A. *Colloids Surf. A* **2000**, *174*, 23-26.
22. Mikula, R. J. *J. Colloid Interface Sci.* **1988**, *121*, 273-277.
23. Povey, M. J. W. *Ultrasonic Techniques for Fluids Characterization;* Academic Press: San Diego, CA, 1997.
24. McClements, D. J. *Adv. Colloid Interface Sci.* **1991**, *37*, 33-72.
25. McClements, D. J. *Langmuir* **1996**, *12*, 3454-3461.
26. Li, D. H.; Ganczarczyk, J. J. *Water Pollut. Res. J. Can.* **1986**, *21*, 130-140.
27. Ríos, G.; Benito, J. M.; Pazos, C.; Coca, J. *J. Dispersion Sci. Technol.* **2002**, *23*, 721-728.
28. Orr, C. In *Encyclopedia of Emulsion Technology;* Becher, P., Ed.; Marcel Dekker: New York, 1983; Vol. 1, pp 369-404.
29. Bueno, J. L.; Dizy, J. *J. Powder Bulk Sol. Tech.* **1987**, *11*, 1-8.

Chapter 6

Room Temperature Laser Light Scattering Determination of the Size and Molecular Weight of Polyamide-11

N. Jones, A. Meyer, S. Lyle, M. Clark, and D. Kranbuehl

Departments of Chemistry and Applied Science, College of William and Mary, Williamsburg, VA 23187

Abstract. This paper describes the development of techniques for making room temperature measurements of the size, molecular weight and distribution in chain lengths of polyamide-11 using multi angle laser light scattering detection with room temperature size exclusion gel chromatography. It is shown to provide information about the amount and length of the longest chains, the critical factor in monitoring the approach to the ductile –brittle transition as the polyamide chains degrade due to hydrolysis during use in the field. Hexafluoroisoproponal is shown to be a good solvent. The addition of salt does not appear to affect the measured value of the molecular weight as expected. Nor does it appear to decrease the size of the chains as the values of the log (Molecular Weight) versus log (radius of gyration) for both the no salt and salt solutions appear to lie along the same line with a slope of 0.55.

Introduction

Development and use of techniques to measure molecular size and conformation in solution remains an essential but often difficult step in characterizing hard to dissolve polymer systems such as polyamides. The appropriate choice and implementation of solvent, measurement conditions and test conditions is a challenging task, given the limited information available when characterizing a polyamide.

The objective of this study is to develop a room temperature procedure to determine the molecular weight, and the size-conformational behavior of polyamide-11, PA-11 in hexafluoroisopropanol, HFIP. A multi angle laser light scattering detector, MALLS was used with size exclusion chromatography,SEC. The role of an additive salt to electrically balance the solvated polymer and potentially reduce solvent chain expansion behavior of the polymer in solution is examined.

Polyamides are not soluble at room temperature in most conventional solvents used for size exclusion chromatography but are soluble at room temperature in hexafluoroisopropanol, HFIP. Thus background studies were carried out on poly(methylmethacrylate), PMMA, standards to verify the compatibility of the HFIP-SEC macromolecular characterization technique. Then studies were made on PA-11 including include commercial samples of Atochem's Besno P40TL, a plasticized PA-11, extruded into pipe. In addition unplasticized PA-11 made in our laboratory by polymerization of 11-aminoundecanioc acid was characterized.

Experimental

Four samples of varying molecular weight were made. These samples were made in an oven flushed with Argon gas and held at 190°C and under a reduced pressure of 380 torr. In order to control the molecular weight, reaction times were varied: 12 hours for the higher molecular weight samples and 1.5 and 0.75 hours for lower molecular weight samples ($M_w \approx 30,000$ and $15,000$), where M_w is the weight-average molecular weight. PMMA narrow molecular weight standards were purchased from Polymer Laboratories, Inc.

Size exclusion chromatography (SEC) was conducted with 1,1,1,3,3,3-hexafluoroisopropanol as the mobile phase at a flow rate of 0.6mL/min with a JordiGel DVB Mixed Bed HPLC column at 40°C. To obtain absolute measurements[1] of weight average molecular weight M_w and root mean square radius R_g by light scattering, a Wyatt miniDawn Multi Angle Laser Light Scattering Instrument (MALLS) with a Wyatt Optilab 903 Interferometric

Refractometer was used. The detectors used light at 690nm, from a diode laser in the miniDawn and from a filtered light source in the Optilab. This system was operated both with and without salt present in the mobile phase. Each sample was run in the system with no salt present, then the set was run with 0.05M Potasium trifluoroacetate salt (KTFA) in the HFIP mobile phase. The dn/dc values for both the PA-11 polymer samples were kept constant when running in both the salt and no salt mobile phase. The values used for PA-11 and PMMA were 0.335 mL/g and 0.188 mL/g respectively. Solutions of PMMA and PA-11 were made at roughly 5.0 mg/mL concentration. M-cresol was used as the solvent for PA-11 and PMMA solutions. An injection volume of 100uL was used to introduce the sample into the mobile phase of the system.

Results

Poly (methylmethacrylate)

The PMMA results shown in Figure 1 and Table I illustrate the typical unimodal peak obtained and the resultant accuracy of the HFIP system with a MALLS light scattering detector:

As seen in Table I, the measured M_w values are quite close to those provided with the known standards. These results verify that MALLS is capable of absolute measurement of PMMA molecular weight in HFIP over a range of roughly 10^3 to 10^6 Daltons. The measured Rg values indicate that the instrument increases in accuracy above 20nm with a lower limit of 10nm. This can be seen in the error of the Rg values below 20nm. Use of a salt such as KTFA in the mobile phase is clearly not necessary for accurate determination of M_w for PMMA in HFIP.

PA-11

Figure 2 illustrates a representative unimodal peak typical in the MALLS and RI output for PA-11 in our system. As is the case with PMMA, there is good separation using the JordiGel DVB column. There is no evidence of distortion in the shape of SEC chromatogram as has been reported in some studies of polyamides in HFIP due to a "polyelectrolyte" effect, and there is no reason therefore to expect error in the M_w calculations because of distortion in the chromatogram. As seen in Table II, MALLS measurements of PA-11 both synthesized in our laboratory and commercial samples show the same value of M_w in 0.05 KFTA salt and without salt in the HFIP mobile phase. This is most strongly observed for the most precise M_w values of the highest molecular weight samples. Even when there is chain extension as will be discussed later, it is relevant to note the MALLS measurement of M_w is an absolute measurement,

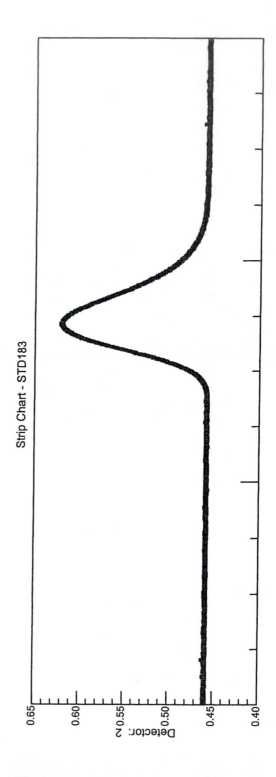

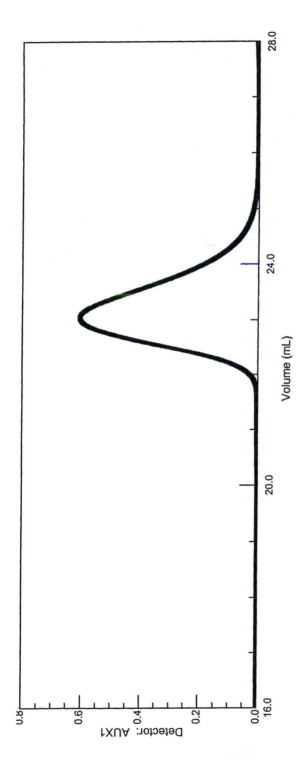

Figure 1: Representative MALLS (light scattering) and RI (concentration) output for PMMA (Mw = 10,000) in the HFIP no salt system.

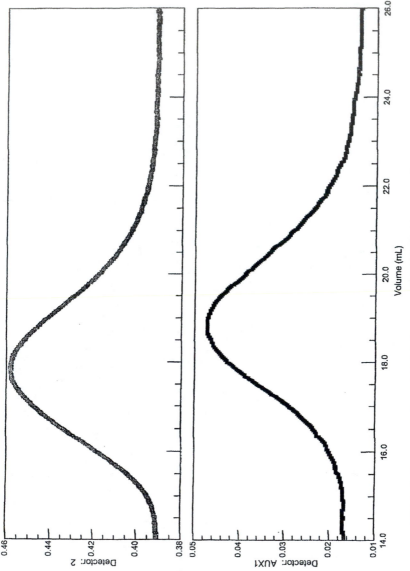

Figure 2: Representative MALLS (light scattering) and RI (concentration) output for commercial PA-11, Mw = 66k, in the HFIP no salt system.

Table I
PMMA Analysis in HFIP No Salt System
Multi Angle Light Scattering

Sample (MW)	Mw	Rg
2400	2080	-
4910	4100	-
6900	6150	-
10000	9744	-
28900	30660	12 + 4
69000	70180	13 + 4
212000	232400	15 + 4
910500	913500	49 + 1

Table II					
MALLS Characterization of PA-11 in HFIP with and without salt					
	SAMPLE	No salt		0.5 M KTFA salted HFIP	
		Mw	Rg	Mw	Rg
A	Unplasticized PA-11 made 6/28/99	17100 ± 1000	12 ± 4	14700 ± 1000	8 ± 4
B	Unplasticized PA-11 made 6/17/99	26010 ± 1200	15 ± 3	24200 ± 1200	17 ± 3
C	Unplasticized PA-11 made 6/3/99	53760 ± 2000	19 ± 2	53300 ± 2000	20 ± 2
D	Commercial PA-11 Asg B	62360 ± 2000	26 ± 2	61400 ± 2000	18 ± 2
E	Unplasticized PA-11 made 6/7/99	66900 ± 2000	17 ± 2	—	—
F	Commercial PA-11 Norsk Hydro	57690 ± 2000	25 ± 2	—	—

regardless of the shape that the dissolved PA-11 molecule assumes in the mobile phase.

Veith and Cohen[2] report a bimodal peak for PA-6 in HFIP without salt due to a "polyelectrolyte" effect. This is not seen for PA-11 in our HFIP system, as demonstrated in Figure 2. In fact, HFIP without salt has been previously demonstrated to be an effective and reliable solvent for PA-11.[3,4] We suggest that the reason for the effect in PA-6 and not in PA-11 is due to the higher number of amide bonds per unit length along the chain backbone in PA-6. The relatively close proximity of the amide linkages in PA-6 gives rise to heightened repulsive forces and the resultant physical expansion of the chain from a relaxed coil to a more expanded form in solution.

Figure 3 shows a representative PA-11 molecular weight data, measured by the SEC-MALLS / HFIP system without salt. As seen in Figure 3, MALLS detection calculates M_w slice by slice as the polymer elutes from chromatography and has a lower limit of approximately 7,500 Daltons for PA-11.

In addition to calculating molecular weights using light scattering measurements, a conventional plot of M_w vs. the retention volume peak using refractive index data from size exclusion chromatography can be constructed, Figure 4. As seen in Figure 4, the accuracy of a calibration using this technique is severely limited by an assumption that the hydrodynamic radii for all of the polymers, both the standards and analytes, are consistent for all given values of M_w. Clearly a major advantage of the MALLS system is that regardless of the drift in the separation column accurate measurements of the molecular weight are made by an absolute and precise multi laser light scattering system. Nevertheless, this technique can be useful in comparing the molecular weight of polydisperse samples, especially those with appreciable amounts of low molecular weight polymer near or below the lower limit of detection by light scattering.

A universal calibration can be used to estimate molecular weights based on a combination of viscosity measurements and retention volumes. However, in HFIP it has been previously and rather conclusively demonstrated that the universal calibration using the product of viscosity and retention volume does not work well for polymers such as polyamides and PMMA in HFIP.[4]

Discussion of Rg

The size and distribution in the lengths of the PA-11 chains is a major focus of this work. The MALLS system directly measures R_g using the angular

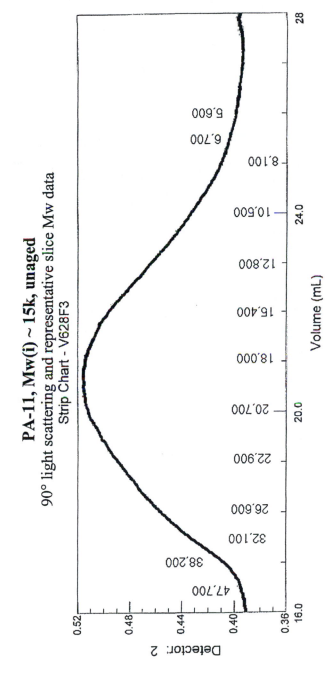

Figure 3: The 90° light scattering signal trace and the molecular weight distribution for PA-11, Mw ~ 15,000 Daltons.

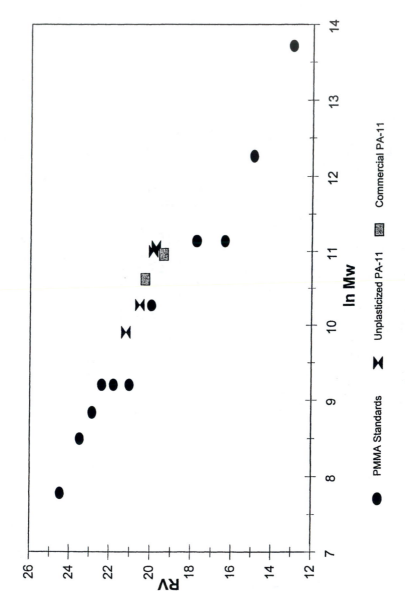

Figure 4: A plot of ln(molecular weight) vs. retention volume for the PMMA and PA-11, without salt

dependence, P_θ, of scattered light from the dissolved polymer. This technique is limited to molecules with Rg of more than about 10nm for reasonable resolution,[5] but the measurement requires no assumption about the shape or conformation of the polymer in solution nor its relation to the hydrodynamic volume. Only knowledge of the exact wavelength of light, λ, and the scattering intensity at each angle θ is used. The relationship between P_θ, λ, and mean square radius is:[6]

$$1 / P_\theta = 1 + (16\Pi^2 / 3\lambda^2) <r_g^2> \sin^2 (\theta/2) \quad (1)$$

Thus because MALLS measures light scattering at multiple angles simultaneously, this method is exact and preferable in cases where the conformation in solution is unknown.

Table II reports values of M_w and the corresponding value of Rg when measured in HFIP with and without salt.

Table II shows that the MALL-SEC values of the weight average molecular weight within the uncertainty are as expected not affected by the presence of the salt as the higher molecular weight values, which are the most precise, are in close agreement. Figure 5 is a plot of the radius of gyration versus Mw for solutions with and without salt. It shows the relation of the radius gyration to Mw and the effect of the presence of salt. Figure 5 shows that a best fit line through a plot of log Rg versus log M_w gives a slope of 0.55 ± 0.09 for all the salt and no salt data. When a line is fit to only the salt data the slope is 0.60 and for only the no salt data the slope is 0.51. Normally the addition of salt, if it has an effect, would be to reduce the extended dimensions of the PA-11 when solvated by HFIP. In this case the addition of salt produces a larger slope and a value within the uncertainty of the slope for the combined salt, no salt data. Thus we conclude addition of 0.05 molar KTFA has no detectable effect on the extended conformation of the polymer chains, indicating the excluded volume effect is unchanged.

Finally we report values of the variation in the lengths of the PA-11 chains for un-aged commercial PA-11, the gas-oil liner used in flexible pipes to transport oil from the ocean floor to a platform. We also show the effect of aging due to hydrolysis of a PA-11 liner after is has been in use for various lengths of time. The PA-11 was initially exposed to the acidic water/oil mixture in the mid-1990s, the results clearly show the difference from the fresh, unexposed PA-11 to the most exposed PA-11 in 2002. These results are shown in Figure 6. Table III reports the changing values of M_w and Rg. These results clearly point to the ability of MALLS, the only currently available direct absolute method, to monitor the molecular weight and distribution in chain lengths of PA-11 as it ages during use in the field.

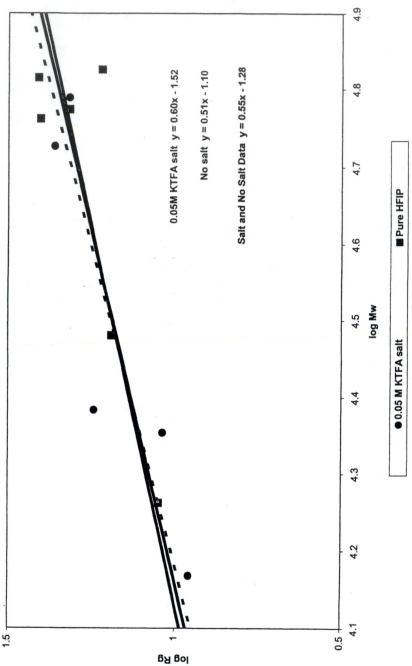

Figure 5: A plot of log Rg vs. log M_w for PA-11 measured using MALLS without salt, with 0.05m KFTA salt .

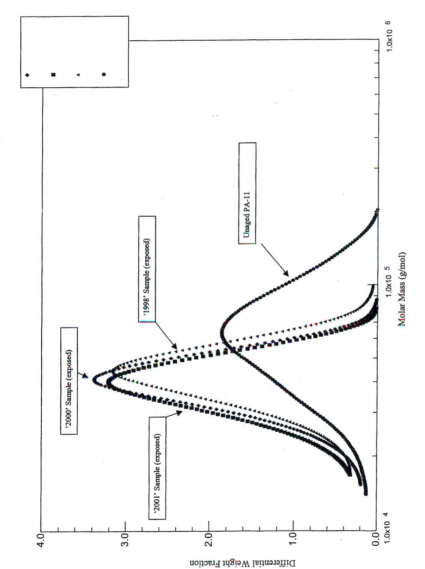

Figure 6: A plot of Differential Weight Fraction *vs.* Molar Mass for several aged and un-aged commercial samples.

Table III

MALLS-SEC Mw and Rg values for aged and un-aged PA-11

Sample	Mw	Rg
Fresh PA-11	65140	22.3
1998 aged PA-11	45150	20.6
2000 aged PA-11	38480	19.3
2001 aged PA-11	37360	17.9

Conclusions

The results for PMMA and PA-11 demonstrate that HFIP is a good room temperature solvent for use in MALLS-SEC measurements of molecular weight for those polymers. The results suggest that there is no need for salt in the solution to characterize the M_w as within the accuracy of the data there is no effect of the presence of salt vs. no salt on the measurement of M_w for the PA-11 in HFIP using MALLS. Within the scatter of the data, the data points for the salt and no salt measurements log M_w vs. log Rg appear to lie along the same line and when taken together the slope is 0.55 ± 0.09 . The linearity and slope of 0.55 suggests that the PA-11 is in an expanded excluded volume state in HFIP and that adding 0.05M KFTA does not significantly change its conformation. The MALLS measurements provide informative plots of the distribution in the size with molecular weight of the chains at or above 7,500 Daltons as well as Rg values above 10nm. Room temperature MALLS-SEC using HFIP appears to be a good absolute method to monitor changes in the PA-11's molecular weight, its size and the distribution in chain lengths resulting from aging hydrolysis mechanisms while in use in the field.

References

1. P. Wyatt. Instrumentation Science and Technology. 25 (1). 1-18. (1997)
2. CA Veith, RE Cohen. Polymer. 30 (5). 942-8. (1989)
3. C. Chuquilin, E. Macchi, R Figini. J. Applied Science. Vol 34. 2433-44. (1987)
4. A. Moroni, T. Havard. Polym. Mater. Sci. Eng. Vol 77. 14-16. (1997)
5. P. Wyatt. Proc. Int. Waters GPC Symp. (1998)
6. Cowie, J. Polymers: Chemistry & Physics of Modern Materials, Second Ed. Blackie Academic & Professional. 196-202. (1993)

Chapter 7

Colloidal Refractometry: The Art of Microstructure Characterization

Mansur S. Mohammadi

Fabric Conditioners, Global Technology Centre, Unilever Research and Development, Port Sunlight, Wirral CH63 3JW, United Kingdom

The knowledge of just two parameters characteristic of colloidal systems can help with the understanding of their microstructure and hence offer insight into their stability, rheology and other performance behaviours. The first is the *in situ* average particle size, which may contain the particle aggregation information, and the second is the volume fraction of the particles. This chapter reviews the emergence of a new structural probe for colloidal liquids in their neat unperturbed state. The technique combines two recent advances in refractometry – an expression for the refractive index of particulate systems as a function of the size and volume concentration of the particles and the availability of digital refractometers which measure the refractive index of concentrated colloids with increased accuracy.

Refractometry, the measurement of refractive index for structural elucidation, has served science since Newton's time. Batsanov's book *(1)* reviews this subject and gives countless examples of refractometric characterisation of homogeneous systems. McGinniss, one of the exponents of the power of refractive index for structural elucidation, states that all physical properties of liquids and solids must be related to their refractive index and gives semi-empirical expressions for such relationships *(2)*. The diverse uses of refractive index in physical chemistry cannot be exaggerated, as a literature search would attest.

This chapter reviews the theoretical and instrumental developments in refractometry of heterogeneous media and the emergence of a generalised refractive index concept which encompasses colloidal liquids. The general refractive index expression contains both the size and the volume concentration of the particulate phase and hence lends itself to the measurement of these sought-after parameters.

Historical Background

Refractive Index of Binary Solutions (Homogeneous Transparent Systems)

The refractive index of a mixture of two miscible transparent liquids or a binary solution of a solvent (v) and a solute (u) can be written as *(3,4)*;

$$n = \phi_v n_v + \phi_u n_u$$

where n, n_v, n_u, represent respectively the refractive indices of the solution, solvent, and solute and ϕ represents the volume fraction. According to this expression the refractive index of a solution, n, is additive in terms of the volume rather than weight concentration of its components. This additivity holds strictly if the volume change upon mixing of solute and solvent is not too large. The Rubber Handbook *(5)* contains many examples of binary solutions for which n is a linear function of the volume fraction but becomes non-linear in terms of the weight fraction of the components. This expression reduces to *(3,4)*;

$$n = n_v + \phi_u (n_u - n_v) \tag{1}$$

in a binary mixture where $\phi_v + \phi_u = 1$. The following two sections show how Equation (1) has evolved to apply to colloidal heterogeneous liquids.

Refractive Index of Colloidal Fluids - Phase Volume Effect

It seems that Wilfried Heller *(6)* made the earliest attempts to derive mixture rules similar to Equation (1) for dispersed systems which included colloidal sulphur, egg albumin, gelatin, silica and titanium dioxide hydrogels. He had realised that for reliable particle size determination by light scattering techniques, based on Mie theory, an accurate value of the refractive index of the particles was necessary. Heller re-wrote Equation (1) for particles (solute) suspended in a medium (solvent) as;

$$n = n_m + \phi_p (n_p - n_m) \qquad (2)$$

and argued that the refractive index of particles, n_p, could be extracted from Equation (2) if the values of n, n_m, and ϕ_p were available. He used polymer latex dispersions as model systems for this purpose and showed that n was indeed linear with ϕ_p. His latices were prepared at infinite dilution to ensure low turbidity and hence acceptable accuracy in the measurement of n (Stokes in his book *(7)* provides a more detailed historical account of the emergence of Equation (2)).

The validity of Equation (2) for realistic volume fractions had to await the systematic investigations of Meeten and co-workers *(8-13)* who developed instrumental techniques to enable the measurement of n for highly turbid dispersions. Figure 1 reproduces a typical set of their results for model monodispersed spherical latices.   Note that, at equal phase volumes, the measured refractive index is smaller for Revacryl (styrene-acrylic copolymer) because of its smaller particle refractive index of $n_p = 1.483$ compared to $n_p = 1.602$ for polystyrene. This is as one expects from Equation (2). Also note the error bars in the refractive index readings of polystyrene latices because of their greater turbidity - a consequence of their larger particle size and refractive index.

Refractive Index of Colloidal Fluids – Particle Size Effect

As early as 1924 Schoorl observed that the refractive index of egg albumin decreased upon coagulation while its viscosity increased *(14)*. Joshi and co-workers found that the refractive index of solutions of hydrophobic colloids decreased during coagulation and used this phenomenon to study coagulation kinetics without understanding the cause of this reduction *(15)*. Deile et al similarly used the changes in refractive index to study the stability effect of methylcellulose on polystyrene latices *(16)*.

In a systematic investigation of the effect of particle size on refractive index of dispersed systems Heller's group *(17)* worked with dilute polymer latices and noticed a sudden drop in n as the particle size increased beyond about 0.40 μm. They were interested in obtaining accurate values for n_p and hence presented their data in terms of an 'effective' refractive index for particles. Figure 2 reproduces their results.

To establish whether this drop in refractive index was caused by some particle chemistry changes or was just a result of size increase Heller and co-workers performed an experiment where, starting with a 0.047 μm size latex, they induced particle aggregation by adding NaCl solution and measured the change in the refractive index with time as reproduced in Figure 3. They concluded that the 'apparent' refractive index of latex particles decreases with their size and ascribed this phenomenon to the enhanced light scattering with increase in particle size.

The explanation of why the refractive index of dispersions decreased with particle size had to await the publication of Van de Hulst's landmark book *(18)* and the investigations of Meeten and co-workers. Following the theoretical foundations laid by Van de Hulst, Meeten et al arrived at;

$$n = n_m + \phi(n_p - n_m)\left(\frac{3\sin p}{p^3} - \frac{3\cos p}{p^2}\right)$$

(3)

where $p = 2\pi d(n_p - n_m)/\lambda_0$, d and λ_0 are the spherical particle diameter and the wavelength of light used in the measurement of n. This equation first appeared in Meeten's book *(19)* for the refractive index of filled polymer melts and in one of his later publications for dispersions *(20)*. A cruder semi-empirical form of Equation (3) has been suggested for particle sizing in liposomal dispersions of detergents and fabric conditioners *(21, 22)*.

This analytical expression for n, which generalises Equation (2), has been made possible by an approximate solution to the Mie light scattering theory called the Anomalous Diffraction Approximation (ADA) *(18, 20, 23)*. ADA applies to a wider range of particle size and m = n_p/n_m than those of Rayleigh and Rayleigh-Gans approximations. Figure 4 compares the exact Mie numerical solution of n with Equation (3) for a typical dispersion. Meeten *(20)* compares ADA with the exact Mie solution of the refraction efficiency parameter for a range of m values for polymer latices. It appears that in most practical situations, where m ≤ 1.3, the exact Mie solution approximates to Equation (3).

Equation (3) explains the experimental observations of Heller and others as typified in Figures 2 and 3 - the refractive index of a dispersion should decrease

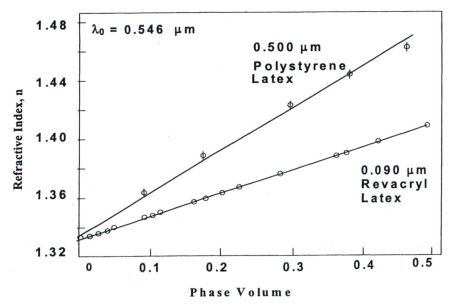

Figure 1. Experimental verification of the linearity of the refractive index with volume fraction for two mono-dispersed latex dispersions. (Reproduced with permission from reference 11. Copyright 1981 Royal Society of Chemistry.)

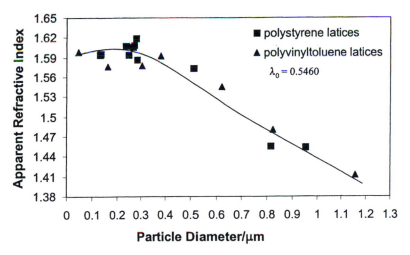

Figure 2. Variation of the apparent refractive index of particles with particle size. (Reproduced with permission from reference 8. Copyright 1977 Elsevier.)

as the particles grow in size. Van de Hulst's book *(18)* provides the physical explanation of this phenomenon..

Building on a brief history of the instrumental developments for measurement of n in colloids, the following sections illustrate the uses of Equation (3) for microstructure characterisation of typical commercial systems.

Measurement of Refractive Index of Dispersions

The accurate measurement of the refractive index of concentrated colloids has not been possible in the past because of their turbid nature. A dark-bright demarcation edge relied on to read the refractive index becomes fuzzy and unreadable for turbid fluids. However recent advances in refractometry have made accurate identification of this cutting edge and hence the accurate measurement of refractive index possible and 'colloidal refractometry' a viable new sizing technique for 'neat' complex fluids.

Reference *(24)* provides a general description of refractometry methods for solutions. Meeten and co-workers *(10-13, 19, 23)* provide an excellent background to the technical development and problems of measuring refractive index for turbid colloidal systems.

Today a variety of inexpensive digital refractometers are available which read the refractive of concentrated dispersions with ± 0.00002 accuracy e.g. Bellingham & Stanley RFM340. The novel feature of these digital machines is the differentiation procedure they apply to the fuzzy light profile found in turbid systems *(22, 25)*. This procedure enables the accurate identification of the lost cutting edge and hence an accurate measurement of the refractive index.

Applications

The inspection of Figure 4 reveals that Equation (3) can apply to two situations:
- When the particle size changes as a result of homogenisation, milling, shearing or in flocculation, aggregation or coalescence processes.
- When the phase volume changes as a result of sedimentation or creaming instability. The refractive index is measured at the interface between the fluid and the refractometer prism. Creaming or sedimentation events change the number density of particles at the interface and hence can be monitored by measuring n with time.

Note that Equation (3) has been derived for monodispersed spherical particles. Practical systems contain polydispersed particle phase or phases.

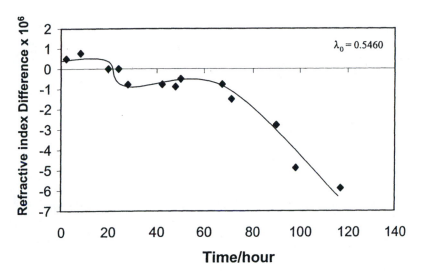

Figure 3. Change in the refractive index of a polyvinyltoluene dispersion with time during the slow aggregation of the particles induced by adding NaCl solution. The refractive index is relative to that at time zero. (Reproduced with permission from reference 8. Copyright 1977 Elsevier.)

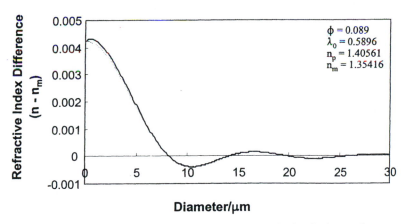

Figure 4. Comparison of the exact Mie numerical solution and Equation (3) for the refractive index of a dispersion of spherical particles (see the inset). Mie is the upper curve.

References *(22, 26, 27)* show that for polydispersed systems with a wide size distribution, polymer latices from 0.2-3.0 μm and liposomal dispersions from 0.1-50 μm, the size that one extracts from Equation (3) approximates $d_{4,3}$, the volume average (see figure 3 in reference *(22)* for more details).

Dairy Products

Table 1 typifies the study of a colloidal system by refractometry.

Table I. Comparison of n and fat droplet size $d_{4,3}$ in μm for off-the-shelf milk samples

Milk	n	$d_{4,3}$ Eq. (3)	$d_{4,3}$ Mastersizer
Skimmed (homogenised)	1.34872	-	-
Semi-skimmed (homogenised.)	1.35001	1.58	0.46
Full fat (homogenised)	1.35131	1.68	0.71
Full fat (unhomogenised)	1.34784	4.92	3.38

Note: All n values were measured at 20 °C and λ_o = 0.5896 μm using B&S RFM340 refractometer. For demineralised water it reads 1.33300. The Mastersizer uses the Mie light scattering theory to calculate scattering intensities for particles of known size and refractive index and extracts size values by matching these calculated intensities with the measured ones. It requires a presentation value (n_p) for the particles from its fixed data set for which 1.4564 was selected as the nearest value to n_p = 1.4566 of fat droplets. The size values from Equation (3) correspond to a match between the measured n value and the calculated n as a function of size.

Figure 5 plots Equation (3) for fat droplets suspended in a milk base and helps to make sense of the n readings in Table 1. For the equation parameters we have used n_m = 1.34872, as measured here for the fat-free skimmed milk, and typical literature values *(28-30)* of 3.5% (full-fat) and 1.7% (semi-skimmed) for the phase volumes and 1.4566 for n_p. According to Figure 5 fat globules between 4 to 7 μm should contribute negatively to the refractive index of milk which explains why n for the unhomogenised milk is less than that of skimmed milk.

For the same system parameters Equation (3) estimates the average fat droplet sizes that appear in the table. Light microscopy ruled out droplet flocculation as the cause of disagreement between the Mastersizer and Equation (3). The literature borrowed system parameters for these milk samples which have different sources may explain the mismatch between the two techniques. However it should be noted that the sizes from Equation (3) seem self-consistent.

One does not expect large size differences between the homogenised semi-skimmed and full-fat samples as shown by the Mastersizer.

Fast Forecasting of Sedimentation and Creaming

Another way to use the colloidal refractometery involves monitoring the changes in ϕ. The refractive index is measured over a microscopic volume of the sample residing on the refractometer prism. Slow particle movements which take days or months to manifest themselves as visible creaming or sedimentation can be 'seen' in minutes or hours by measuring refractive index with time *(26)*. Sedimentation or creaming instability should lead respectively to an increase or a decrease in the phase volume of particles over the prism-sample interface.

Figure 6 illustrates this type of application of colloidal refractometry for the full-fat milk samples both of which creamed while standing on the refractometer prism. For unhomogenised milk this meant the displacement of larger particles with small contributing ones and hence an increase in n with time whereas for homogenised milk it meant loss of contributing particles, because of their small sizes, and hence a decrease in n with time. Interesting microstructural insight emerges from further analysis of such kinetics data *(26)*.

Liposome Dispersions

Table 2 compares the sizing values for a fabric washing liquid system comprising lamellar 'onions'. References *(31, 32)* explain how n_p, n_m and ϕ were obtained for this system to enable sizing by Equation (3). The agreement between different sizing techniques meant that the refractive index measurement on its own could be used for monitoring size differences in production lines.

Table II. Comparison of $d_{4,3}$ in μm for liquid detergents with liposomes

Liquid	electron microscopy	Mastersizer	Equation (3)
A	3.06	6.54	4.72
B	4.42	5.59	4.92
C	2.99	4.40	3.90
D	2.29	3.33	2.19

Note: The $d_{4,3}$ values of electron microscopy method were calculated using the micrographs to read the particle diameter and the number of particles found in that diameter range. Sample A had some larger than 20 μm droplets that were partly visible in the micrographs and hence not fully counted

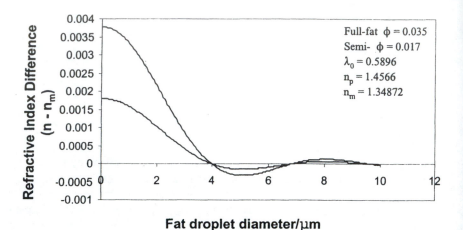

Figure 5. The calculated refractive index of full-fat (upper curve) and semi-skimmed (lower curve) milk samples as a function fat droplet size according to Equation (3).

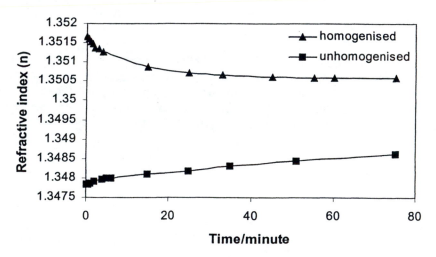

Figure 6. Kinetics of creaming for the full-fat milk samples studied by measuring the change in refractive index with time.

Table 3 compares similar results for (di-chain) cationic liposome dispersions. System A was colloidally more stable than B because of its larger level of a deflocculating polymer. Light microscopy showed both samples to be flocculated. The Mastersizer gave the primary particle size of the flocs because of the dilution stage it required. Homogenisation led to (primary) size reduction for both samples, as reflected in the Mastersizer values, but also led to shear induced flocculation for the less stable sample (B). The sizes from Equation (3) reflect this fact as confirmed by microscopy, rheology, and the way sample B phase separated on standing (26).

Table III. Comparison of $d_{4,3}$ in μm for liquid fabric softeners with liposomes

Liquid	Mastersizer	Equation (3)
A	4.8	9.0
A + homogenisation	1.8	8.9
B	6.3	8.7
B + homogenisation	2.7	15.0

Note: see (26) for electron micrographs of these samples.

Macro-emulsions

Table 4 illustrates another example of sizing in a silicone oil in water macro-emulsion. Using the system parameters given in the footnote to the table, Equation (3) shows that $(n - n_m)$ becomes zero at about 8.15 μm. Therefore droplets larger than this size will fall on the oscillatory region of size-n curve for where more than one size can match a given value of n as the examination of a typical n-size curve reveals (see Figure 4). No size that matched the measured refractive indices could be found for emulsions 1-5 in Table 4.

Table IV Comparison of $d_{4,3}$ in μm for silicone oil in water macro-emulsions

Emulsion	Measured n	Mastersizer	Equation (3)
1	1.35374	129	no match
2	1.35376	137	no match
3	1.35359	88	no match
4	1.35374	52	no match
5	1.35355	24	no match
6	1.35407	22	25.5
7	1.35425	9	14.1

Note: The droplet phase (silicone oil and surfactant) $n_p = 1.40561$, the aqueous phase (a glycerol solution) $n_m = 1.35416$, and $\phi_{droplets} = 0.0827$. Only emulsion 7 reads a refractive index that exceeds n_m where the particles contribute to n.

Limitations

The above example highlights one of the limitations of the colloidal refractometry for sizing – no meaningful sizing results for size which fall within the oscillatory region or outside the measuring range. Nevertheless the information could be useful in itself e.g. confirming the existence of large particles or in combination with microscopy and other sizing techniques.

Another limitation relates to the fact that Equation (3) is derived for spherical particles. This limitation is common to all light scattering techniques based on the Mie theory and renders the sizing information of comparative rather than absolute value for non-spherical particles. Non-spherical particles can be evaluated as the equivalent distributions of spherical particles. Ellipsoids can be presented by a certain size distribution of spheres, for example *(33-35)*. In polydispersed systems where particles span a few decades in size shape correction is not that important to the size measurement.

Future Vistas

The modern digital refractometers can be programmed to take input data on n_m, ϕ and n_p and to read out $d_{4,3}$ instead of n using Equation (3). Measurement of n at larger wavelengths allows the sizing range of Equation (3) to be extended *(36)*. By measuring n at a number of wavelengths and solving the simultaneous equations of n the values of n_m, n_p and ϕ can be obtained as has been achieved in turbidometry (37).

Conclusions

The concept of refractive index has been the corner stone of microstructural understanding in solutions and homogeneous systems in the past. This chapter presented how this traditional concept has evolved during the last two decades to encompass colloidal heterogeneous liquids. An inexpensive modern digital refracotometer can measure the refractive index of concentrated emulsions or dispersions with considerable accuracy. The generalised expression then provides the insight into the microstructural meaning of this measurement as the examples in this chapter illustrated

No single measurement technique can on its own fully characterise the structure of complex fluids. The colloidal refractometry presented here becomes a potent probe for microstructural elucidation in combination with other sizing techniques.

References

1. Batsanov, S.S. Refractometry and Chemical Structure,Van Nostrand, Princeton, NJ, 1966
2. McGinnis, V.D. Org. Coat Plast. Chem. **1978**, 39, 529-534
3. Bottcher, J.F. The Theory of Dielectric Polarisation, Elsevier, 1952
4. Glover, F.A.; Goulden, J.D.S. Nature, **1963**, 200, 1165
5. Handbook of Chemistry and physics, Lide, D.E., CRC Press, Boca Raton, Fl, 2001-2002, 82nd edition, pages 8-57 to 8-83.
6. Heller, W. Physical Review, **1945**, 63(1,2), 5-10
7. Stokes, A.R. The Theory of the Optical Properties of Inhomogeneous Materials, E. and F.N. Spon Limited, London, 1963
8. Champion, J.V.; Meeten, G.H.; Senior, M. J. Chem. Soc. Faraday Trans. 2 ,**1977**, 74(7), 1319-1329
9. Champion, J.V.; Meeten, G.H.; Senior, M. J. Colloid Interface Sci, **1979**, 72 (3)
10. Killey, A.; Meeten, G.H. J. Chem. Soc. Faraday Trans. 2, **1981**, 77, 587-599
11. Alexander, K.; Killey, A.; Meeten, G.H; Senior, M. J Chem. Soc. Faraday Trans. 2, **1981**, 77, 361-372
12. Meeten, G.H.; North, A.N. Measu. Sci. Technol. **1991**, 2, 441-447
13. Meeten, G.H.; North, A.N. Measu. Sci. Technol. **1995**, 6, 214-221
14. Schoorl, N. Rec. Trav. Chim. **1924,** 43, 205
15. Joshi, S.S.; and others J. Indian Chem. Soc. **1936**, 13, 141, 217, 309
16. Deilie, O.; Kramer, H. Klause, W. Kolloid-Z. **1953**, 130, 105
17. Heller, W.; Pugh, T.L. J. Colloid Science **1957**, 12, 294-307
18. Van de Hulst, H.C. Light Scattering by Small Particles, Dover Publications, New York, 1981 (1957 Wiley and Son publication)
19. Meeten, G.H. in Optical Properties of Polymers, Meeten, G.H., Ed., Elsevier, London, 1986, chapters 1 and 6
20. Meeten, G.H. Optical Communications, **1997**, 134, 233-240
21. Buytenhek, C.J.; Mohammadi, M.S.; van de Pas, J. C.; Schepers, F.J.; De Vries, C.L. patent WO 91/09107, 1991
22. Mohammadi, M.S. Advances in Colloid and Interface Science, **1995**, 62, 17-29
23. North, A.N. Ph.D. Thesis, City of London Polytechnique, London, 1986
24. Stanley, G.F. Refractometers: Basic Principles, Bellingham & Stanley Ltd, 1989
25. Geeke, J.E.; Mill, C.S.; Mohammadi, M.S. Measu. Sci. Technol. **1994**, 5, 531-539
26. Mohammadi, M.S. J. Dispersion Science and Technology, **2002**, 23(5), 689-697

27. Mohammadi, M.S. a poster presentation, European Colloid and Interface Society, 22-27 September, 2002
28. Walstra, P. Neth. Milk & Diary J., **1965**, 19, 93
29. Walstra, P.; Borggreve, G.J. Neth. Milk & Diary J., **1966**, 20, 140
30. Jääskeläinen, A.J. ; and others J. Dairy Sci. **2001**, 84, 38-43
31. Van de Pas, J.C.; and others Colloids surfaces A. Phys-Chem Eng. Aspects, **1994**, 85, 221-236
32. Kevelam, J.; and others Langmuir **1999**, 15, 4959-5001
33. Glatter, O; and others Part. Part. Syst. Charact., **1991**, 8, 272-281
34. Glatter, O.; Hofer, E. J. Colloid & Interface Science, **1988**, 122(2), 484-495
35. Wang, D-S; and others Applied Optics, **1979**, 18(15), 2672-2678
36. Mohammadi, M.S.; Geek,J.E.; Mill, C.S. unpublished work
37. Hosono, H.; and others Bullet. Inst. Chem. Res., Kyoto University **1973**, 51(2), 104-117

Fractionation Methods and Applications: Ultracentrifugation, Capillary Hydrodynamic Fractionation, and Field-Flow Fractionation

Chapter 8

Analysis of Nanoparticles <10 nm by Analytical Ultracentrifugation

Helmut Cölfen

Max-Planck-Institute of Colloids and Interfaces, Colloid Chemistry, Am Mühlenberg 2, D–14476 Golm, Germany

INTRODUCTION

Nanoparticles with sizes < 10 nm became increasingly important throughout the last years either due to their high surface to volume ratio for applications like catalysis or due to their size dependent optical or electrical properties in this size range (quantum size effect). Therefore, an increasing demand exists to analyze such particles with great precision. Nevertheless, many of the existing techniques for particle size analysis are not particularly well suited for such small particles. For example, light scattering is problematic as smallest nanoparticles scatter light only to a small extent so that traces of larger impurities are a problem. Analytical ultracentrifugation (AUC) on the other hand is a technique particularly well suited for particles in this size range as the high centrifugal fields of up to 290,000 g allow the fractionation of mixtures so that true particle size distributions can be obtained and the optical detection systems detect every sedimenting particle so that statistical problems as for microscopic techniques do not exist. In addition, microscopic techniques mostly require a drying step which can produce artifacts according to aggregation, selective surface adsorption etc.

AUC is a powerful fractionating technique for polymer and particle characterization and has played a significant part in the understanding of colloidal but especially macromolecular systems starting with the pioneering work of Svedberg (1, 2) who initially invented this technique for the characterization of particle sizes (1, 3). AUC is an absolute method requiring no standards and covering a very broad application range of molar masses between 200 and 10^{14} g/mol and particle sizes between <1 and 5000 nm. The power of the technique lies in the fractionation of the sample into its components either according to their molar masses/particle sizes or to their structure/density without the need of any stationary phase as required in many chromatographic methods. The fractionation enables the measurement of distributions of molar masses, particle sizes and densities and different types of basic experiments with the same instrument can yield complimentary physicochemical information. Often, AUC is the only applicable technique if complex mixtures are to be investigated. There exist no limitations in the choice of solvents even at extreme pH. This results in a whole variety of applications and although the technique now already exists for about 80 years, there are still new methods and applications coming up.

In this paper, the range and quality of information which can be obtained by AUC of nanoparticles < 10 nm will be presented using examples from the authors laboratory.

INSTRUMENTATION

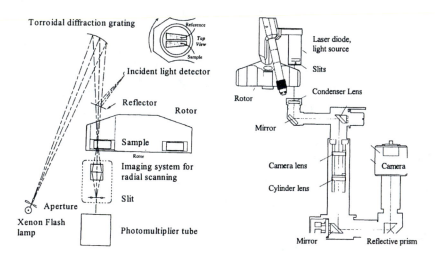

Figure 1: Optical detection systems of the Optima XL-I analytical ultracentrifuge. Left: UV/VIS-absorption optics, Right: Rayleigh interference optics based on the construction of Laue (4). Figures reproduced with kind permission of Beckman Coulter, Palo Alto, CA.

An analytical ultracentrifuge is an ultracentrifuge with one or several optical detection systems which allow the observation of a sedimenting sample in a centrifugal field of up to the 290.000 fold gravitational force (see Fig. 1). The sample is placed in a single or double sector cell (see Fig. 1) where the sector shape prevents convection during sedimentation due to the radially sedimenting sample. In addition to the two optical systems modern machines are equipped with (see Fig. 1) and which detect the radially changing concentration in the cell, the Schlieren optical system which detects the refractive index/concentration gradient can be applied (5). It delivers the first derivative of the radial concentration gradient and is thus well suited to watch sedimenting boundaries during sedimentation velocity experiments. Probably, the typical Schlieren peak is the most well known experimental output of an analytical ultracentrifuge. All three optical systems have their special advantages. The UV-absorption optics combines sensitivity with selectivity due to the variable detection wavelength whereas the Rayleigh interference optics yields very accurate experimental data due to the acquisition of a number of interference fringes which are then evaluated via a fast Fourier transformation. However, the interference optics can only determine relative concentration changes with respect to a fixed point which is usually the meniscus air/solution. The Schlieren optics is well suited for high concentration or density gradient work and all kinds of experiments where a derivative of the concentration gradient is needed for the evaluation (e.g. the determination of the z-average molar mass from sedimentation equilibrium). The Schlieren optics is similar to the setup of the Rayleigh interferometer in Fig. 1 but has a phaseplate or knife edge in the focus of the condenser lens as an additional element.

The typical optical patterns derived by the three mentioned detection systems are shown in Fig. 2. For an extensive description of these optical systems see Ref. (5). For modern ultracentrifuges, only the Rayleigh interferometer (as an on-line detector) and the absorption optics are still available although the Schlieren optics was recently adapted for the use in the Beckman Optima XL/XL-I (6). As further detection systems, a fluorescence detector (7) as well as a turbidity detector specially designed for particle size analysis mainly of latices (8-11) were reported.

It is generally advantageous to combine several optical systems. Especially the combination of the Rayleigh interference optics and the UV/Vis absorption optics can yield important informations about complex systems where for example an absorbing component is selectively detected with the absorption optics whereas the Rayleigh interferometer detects all components.

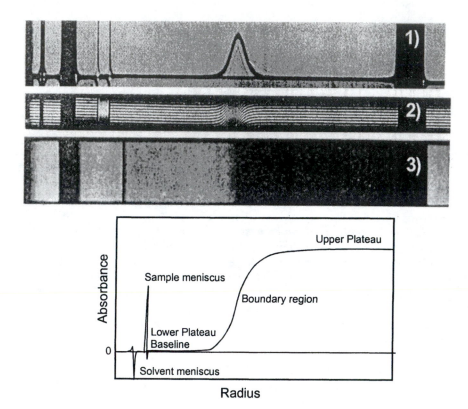

Figure 2: Photographically recorded optical data obtained from the different detection optics of an analytical ultracentrifuge for the same sample. 1) Schlieren optics, 2) Rayleigh interference optics and 3) absorption optics. The lower picture presents an output of a scanning absorption optics. Figure partly reproduced from Ref. (12) with permission of Elsevier and Beckman Instruments.

EXPERIMENT TYPES

To obtain a particle size distribution by AUC, a so-called sedimentation velocity experiment is performed. As this paper focuses on the analysis of nanoparticles < 10 nm which are predominately of inorganic nature, the other AUC experiment types (Sedimentation equilibrium, density gradient and synthetic boundary experiment) are not treated here although they may also prove useful for the analysis of very small nanoparticles. For a more detailed treatment of these AUC methods for nanoparticles, the reader should consult Ref. (13).

Sedimentation velocity experiment:

A sedimentation velocity experiment is carried out at high centrifugal fields and is the most important AUC technique for nanoparticle characterization. Here, the molecules/particles sediment according to their mass/size, density and shape without significant back diffusion according to the generated concentration gradient. Under such conditions, a separation of mixture components takes place and one can detect a step-like concentration profile in the ultracentrifuge cell, which is broadened with time due to diffusion. The profile usually exhibits an upper and a lower plateau. Each step corresponds to one species. If one detects the radial concentration gradient in certain time intervals, the sedimentation of the molecules/particles can be monitored (Fig. 3a).

From the velocity of the sedimenting boundary, one can determine the weight average sedimentation coefficient s_w in a classical way according to (Fig. 3b):

$$s = \frac{\ln(r/r_m)}{\omega^2 t} \tag{1}$$

where r with respect to the rotational axis is the position of the midpoint or second moment point of the moving boundary, r_m the radial distance of the meniscus, t the time and ω the angular velocity of the rotor. The sedimentation coefficient is a concentration and pressure dependent quantity which can be taken into account by appropriate correction or the extrapolation of a concentration series to zero concentration. However, often small nanoparticles are coloured which permits their analysis at very low concentrations with the UV/Vis absorption optics so that extrapolations to infinite dilution are not necessary. On the other hand, an intense colour restricts the use of interference optics as it smears the fringes.

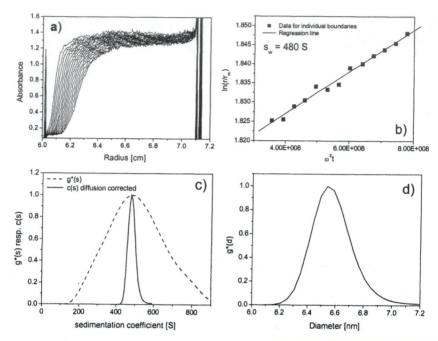

Figure 3: Sedimentation velocity experiment on gold colloids in water at 5000 RPM and 25 °C illustrating various evaluation methods. a) Experimental raw data acquired with scanning absorption optics at 575 nm. Scan interval 2 min. Radius means the radial distance to the center of rotation. b) Sedimentation coefficient calculated from eq. (1). c) Apparent sedimentation coefficient distribution g*(s) from the time derivative method eq. (3) and Ref. (14) as well as diffusion corrected sedimentation coefficient distribution c(s) (39). d) Resulting diffusion corrected particle size distribution calculated from eq. (5)

A plot of $\ln(r/r_m)$ vs. $\omega^2 t$ is a line with the slope s = sedimentation coefficient (see Fig. 3b). The sedimentation coefficient is measured in the unit Svedberg (S) where $1\ S = 10^{-13}$ s. If the diffusion coefficient D is known from other experiments (light scattering, ultracentrifugation etc.), one can calculate the molar mass of the sample according to the Svedberg equation:

$$M = \frac{sRT}{D(1 - \bar{v}\rho)} \qquad (2)$$

where M is the molar mass of the sample, R is the gas constant, T the thermodynamic temperature, $\bar{v}$ the partial specific volume and ρ the solvent density. The partial specific volume is a critical value, but it can be determined with good precision in a density oscillation tube if enough sample material is available. For hybrid particles in mixtures with different compositions, $\bar{v}$ can not be measured as only the average over all particles in the mixture is obtained. Here, the workaround of reasonable assumptions must be applied or quantitative density gradient experiments can yield the average density from the density distribution. Density gradients are however not possible for inorganic particles as their density is too high so that the determination of the density of individual particles in a complicated mixture is still a problem.

Van Holde Weischet method:

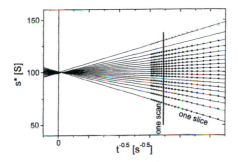

Figure 4: Typical van Holde-Weischet Plot of a sedimentation velocity experiment with a monodisperse system. Reproduced from Ref. (15) with permission.

If nanoparticles < 10 nm are investigated by AUC, diffusion broadening of the sedimenting boundaries will significantly occur due to the high diffusion coefficients of the particles (see Fig. 3c).. An approach to remove the effects of boundary spreading by diffusion and thus enabling to calculate the diffusion corrected integral s-distribution G(s) was introduced by van Holde and Weischet (16). This is done by selecting a fixed number of data points from one experimental scan that are evenly spaced between the baseline and the plateau. Then, an apparent sedimentation coefficient s^* is calculated for each of the data points and plotted versus the inverse root of the runtime yielding the typical van Holde-Weischet Plot (See Fig. 4).

If a linear fit of the corresponding $s*$ (one slice) is performed, the integral diffusion corrected sedimentation coefficient distribution G(s) can be obtained from the y-values at infinite time in the van Holde / Weischet plot. In case of a single monodisperse component, the lines intersect in one point (see Fig. 4). For multiple components, the corresponding number of intersects is obtained whereas the intersection point is shifted to times less than infinity in case of nonideality. Therefore, the van Holde/Weischet analysis is a rigorous test for sample homogeneity or nonideality (16-21). However, it has to be mentioned that the theory behind the van-Holde/Weischet method is not valid for very small particles with strong diffusion so that the boundary smearing of several nanoparticles to a single boundary cannot be resolved (22).

Time derivative method:

In many cases, particles are polydisperse or one detects a multimodal distribution. In such cases, it is of interest to determine the sedimentation coefficient distribution $G(s)$ or the differential form thereof $g(s)$. Although this is in principle possible by the van Holde Weischet method, a better suited method for the determination of $g(s)$ is the time derivative method (14 23, 24) which determines the time derivative of the radial scans acquired at different times according to

$$g*(s)_t = \left(\frac{\partial\{c(r,t)/c_0\}}{\partial t} \right) \left(\frac{\omega^2 t^2}{\ln(r_m/r)} \right) \left(\frac{r}{r_m} \right)^2 \qquad (3)$$

with c = concentration. $g*(s)$ equals the true distribution $g(s)$ in cases where diffusion can be ignored. An example for gold colloids is given in Fig. 3c. An important advantage of this procedure is a significant improvement of the quality of the calculated distribution because two scans are subtracted from each other so that systematic errors in the optical patterns (baseline) cancel out and the random noise decreases. However, a drawback of the time derivative method is that only scans from a relatively narrow time interval can be used for a single evaluation so that in fact, no full advantage is taken of the possibility to scan several hundreds of experimental scans throughout an experiment. If diffusion is significant, extrapolation of $g(s*)$ curves calculated for different times to infinite time yield the true distribution.

The determination of g(s*) can already yield a lot of important information beside the sample homogeneity and number of components. In case of interacting systems for example, interaction constants can be derived (25). If the reaction rates are slow compared to the experimental time, multiple Gauss curves with a maximum at fixed s can be fitted to g(s) – each of them corresponds to one component (26). By that means, the aggregation state as well as the corresponding concentration of the different aggregates can be determined for aggregating systems so that the equilibrium constant and thus the free enthalpy of the association steps is accessible. An example of the advantageous application of this technique was reported for the precrystallization aggregation of lysozyme which yielded the smallest oligomer being able to form a crystal (27).

Fitting to approximate or finite element solutions of the Lamm equation:

One new and recently much investigated approach has proved useful for the determination of s and D and thus M as well as the concentrations of individual components from sedimentation velocity data like in Fig. 3a. This is the fit of a series of radial concentration profiles to approximate solutions of the Lamm (28) differential equation of the ultracentrifuge (29-34) or finite element solutions of the Lamm equation (35-38).

$$\frac{\partial c}{\partial t} = -\frac{1}{r}\frac{d}{dr}\left(\underbrace{r\,D\frac{dc}{dr}}_{\text{Diffusion term}} - \underbrace{s\,\omega^2 r^2 c}_{\text{Sedimentation term}} \right) \tag{4}$$

As the Lamm equation (4) is the fundamental equation in analytical ultracentrifugation capable to describe all types of ultracentrifuge experiments, fitting of experimental data to this equation is a potentially very powerful approach. However, a drawback is that this method is clearly model dependent.

Very recently, the requirement of monodisperse samples for fitting of the Lamm equation was overcome so that now even sedimentation coefficient and molar mass distributions of polydisperse samples can be investigated (39). However, for the determination of molar mass distributions; this approach suffers from the necessary prior knowledge of $\bar{v}$ and the frictional ratio f/f$_0$ of the sample (f is the frictional coefficient and f_0 is the frictional coefficient of the spherical particle having the same mass and $\bar{v}$ as the sample under consideration). If the frictional ratio is not known, it can also be fitted which in turn allows conclusions about the particle shape.

The most significant merit of this approach for nanoparticle analysis is the possibility to correct for the effects of diffusion on the broadening of the sedimenting boundary so that diffusion corrected sedimentation coefficient distributions can be obtained reflecting the true sample polydispersity. This is clearly shown in Fig. 3c. The method was recently shown to yield reliable results for different model systems (22).

Comparison of sedimentation velocity evaluation methods:

The above presented methods have their special merits and are thus briefly compared to allow an assessment of the methods. The vanHolde / Weischet method is a rigorous test for sample homogeneity and nonideality and allows to correct for diffusion broadening of sedimenting boundaries. However, as the theory behind it is not valid for very small particles with strong diffusion, it is only of a very limited value for the analysis of nanoparticles. The time derivative method is valuable, if distributions of a good quality are desired but it does not allow for diffusion correction of the distribution. Another drawback is, that only scans of a limited time interval can be analyzed so that this methods does not make use of the possibility of modern ultracentrifuges to acquire hundreds of experimental scans in a single experiment, unless scans are evaluated for several time intervals. In that case, the calculation of the diffusion coefficient is possible. Fitting to approximate solutions of the Lamm equation can make full use of the possibility to acquire large experimental datasets and thus yields very accurate results. In addition, time and radially invariant noise can be removed and it is possible to obtain diffusion corrected sedimentation coefficient distributions - even for polydisperse particles. However, in case of polydispersity, artificial peaks can be generated by the diffusion correction so that it is a good advice to compare the diffusion corrected with the uncorrected distribution.

PARTICLE SIZE DISTRIBUTIONS

The application of AUC for the determination of particle sizes and their distributions to address problems of colloid analysis was already realized by the pioneers of this technique because sedimentation velocity experiments provide a sensitive fractionation due to particle size/molar mass (3, 40-42). Nevertheless it appears, that the potential of this application is still not yet commonly recognized in the colloid community. It is relatively straightforward to convert a sedimentation coefficient distribution which can be calculated using eq. (1) or one of the above evaluation methods for every data point r_i if a radial scan has been acquired at a specified time to a particle size distribution. Alternatively, diffusion corrected sedimentation coefficient distributions calculated according to Ref. (39) prove of great value especially for colloids < 10 nm as evident from Fig. 3c. Assuming the validity of Stoke's law (e.g. the sample is spherical), the following derivative of the Svedberg equation eq. (2) is obtained

$$d_i = \sqrt{\frac{18\eta s_i}{\rho_2 - \rho}} \tag{5}$$

where d_i is the particle diameter corresponding to s_i and ρ_2 is the density of the *sedimenting* particle (including solvent/polymer etc. adhering to the sample) and η the solvent viscosity. This directly allows the conversion of a sedimentation coefficient distribution to a particle size distribution. If the particles are not spherical, only the hydrodynamically equivalent diameter is obtained unless form factors are applied if the axial ratio of the particles is known from other sources like TEM.

The conversion of sedimentation coefficient distributions to a particle size distribution highly relies on the knowledge of the density of the sedimenting particle. For hybrid particles or very small nanoparticles < 10 nm, this issue can be a severe problem, especially in case of mixtures, as the density of the particles is usually not known. Measurements of the average particle density in the mixture can lead to erroneous results so that in such cases, the correlation of the sedimentation coefficient distribution with a distribution obtained from a density insensitive method like Flow-Field-Flow Fractionation or dynamic light scattering is meaningful. This can in turn yield the particle density which can give information about the relative amount of the materials building up the hybrid particle (43). But even an apparent particle size distribution which is calculated within the limits of reasonable particle densities can already yield very valuable information (44). The following examples illustrate the fractionation power of the analytical ultracentrifuge for inorganic colloids.

In case of inorganic colloids, AUC proves to separate dispersions with an almost atomar resolution. The most striking example is shown in Fig. 5a, where Pt was quenched during the growth process and all species in an eight component mixture were successfully analyzed with Angström baseline resolution with the smallest compound being just a cluster of 21 Pt atoms and the shown particle size distribution allows the conclusion that the particles grow via coalescence of primary particles (45). HRTEM only yielded an overall bimodal distribution with the same sizes as in Fig. 5a suffering from the lack of statistical

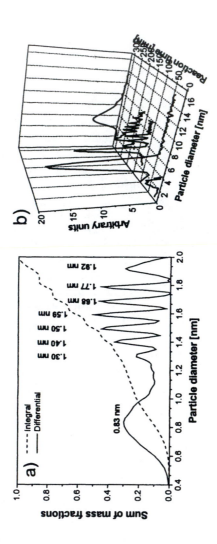

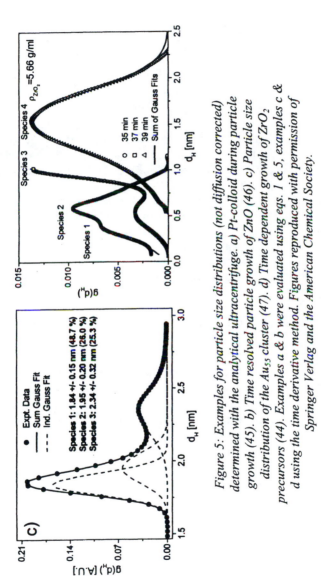

Figure 5: Examples for particle size distributions (not diffusion corrected) determined with the analytical ultracentrifuge. a) Pt-colloid during particle growth (45). b) Time resolved particle growth of ZnO (46). c) Particle size distribution of the Au_{55} cluster (47). d) Time dependent growth of ZrO_2 precursors (44). Examples a & b were evaluated using eqs. 1 & 5, examples c & d using the time derivative method. Figures reproduced with permission of Springer Verlag and the American Chemical Society.

significance in contrast to AUC where all particles are detected (data not shown). The smallest reported species which could be detected so far was for a ZrO_2 precursor and just 0.4 nm in diameter (44) (Fig. 5d) which shows the potential of the analytical ultracentrifuge for the analysis of smallest nanoparticles or subcritical complexes and their growth. This means that particle size distributions derived by ultracentrifugation can well be used to investigate particle growth mechanisms (44, 45) from its initial stages / the critical crystal nucleus, especially if the growth is slow enough to detect the time dependent particle growth like in the example of ZnO (see Fig. 5b) (45, 46). However, this is a rare case and if the samples cannot be successfully quenched as shown in Fig. 5 a & d, the ultracentrifuge is usually too slow to pick up kinetic information.

The fractionation power of AUC is also illustrated in Fig. 5c where the Au_{55} cluster was investigated (47) which was reported to be a definite monodisperse species which is certainly not correct and the resolution between the detected species is in the Angström range.

A rapidly importance gaining substance class are hybrid colloids between polymers and inorganic matter. Here, analytical ultracentrifugation shows all its merits for the investigation of transformations, aggregation processes etc. For example, the encapsulation of a molybdenium cluster with a surfactant could be characterized as well as the aggregation of the primary clusters in different solvents with high resolution (48).

OBSERVATION OF CHEMICAL REACTIONS

It is possible to perform chemical reactions in the analytical ultracentrifuge with a synthetic boundary experiment using a so called synthetic boundary cell (49). In this cell, a small amount of a reactant (usually 10-15 µl) is layered onto a column of the second reactant while the centrifuge is speeded up to about 3000 RPM establishing a very defined reaction boundary. The reaction products are then separated by the centrifugal field and detected by the optical systems of the AUC. This type of synthetic boundary experiment was recently adapted to carry out crystallization reactions inside the spinning cell of the ultracentrifuge which was named "Synthetic Boundary Crystallization Ultracentrifugation" (49, 50). Here, a small amount of Na_2S was layered upon a solution of $CdCl_2$ containing a stabilizer. A fast reaction to CdS with subsequent stabilization of the formed very small nanoparticles takes place according to the following scheme (49):

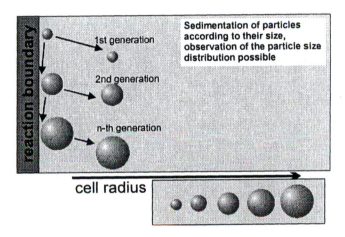

Fig. 6: Schematic representation of Synthetic Boundary Crystallization Ultracentrifugation. Reproduced from Ref. 49 with permission.

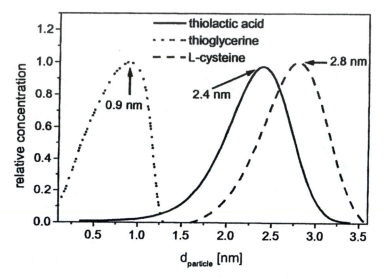

Fig. 7: Apparent particle size distributions of CdS in the presence of different stabilizer molecules as detected by Synthetic Boundary Crystallization Ultracentrifugation and evaluated via eqs. 1 & 5 (49, 50). Reproduced with permission of Elsevier.

The virtue of this technique lies in the fast chemical reaction within the very small reaction zone and the quenching of the further particle growth in lack of the second reaction partner as soon as the particles move out of the reaction zone by sedimentation or diffusion processes. These particles are then fractionated according to their size and density and a particle size distribution can be determined as discussed above. However, as the formed nanoparticles are very small, they show significant diffusion so that they can diffuse back into the reaction boundary and grow further thus forming the second growth generation of particles. This process will be repeated until the reaction partner in the reaction zone will be used up (which usually occurs within a few minutes), so that the particles now only sediment due to their size/density with the usual diffusion broadening of the boundary so that in fact a fast time distribution during a particle growth process is converted to an experimentally accessible radial distribution. By this technique, different growth stages of nanoparticles can be investigated. However, due to the extensive diffusion of the particles, the individual particle size distributions become extensively smeared so that they are only detected as a continuous distribution anymore unless a diffusion correction is performed which only very recently was developed for this sort of experiments (39). Nevertheless, even data not corrected for diffusion can show the different stabilizer capabilities with a resolution in the Angström range as shown for CdS in Fig. 7 (49, 50).

CONCLUSION

Analytical ultracentrifugation is one of the most powerful techniques known to date for the characterization of very small colloids < 10 nm. The power of AUC lies in the fractionation of the sample and therefore the possibility to measure distributions without the interaction with any stationary phase or solvent flows as occurring in the nowadays commonly applied chromatographic techniques. Where ever information is sought for the individual components in a mixture, AUC is the among first techniques of choice and especially for many nanoparticles < 10 nm, the only realistic choice to determine a particle size distribution. Therefore, it is expected that AUC will play an important future role in colloid chemistry if only researchers recognize the tremendous potential of this method.

REFERENCES

1. Svedberg T., Nichols J.B.; *J. Am. Chem. Soc.* **1923**, 45, 2910
2. Svedberg T., Pedersen K.O., *The Ultracentrifuge*, Clarendon Press, Oxford, England, 1940

136

3. Svedberg T., Rinde H., *J. Am. Chem. Soc.* **1924**, 46, 2677
4. Laue T.M. in *Analytical ultracentrifugation in biochemistry and polymer science* Harding S.E., Rowe A.J., Horton J.C., Eds., The royal society of chemistry, Cambridge, England, 1992, 63
5. Lloyd P.H.; *Optical methods in ultracentrifugation, electrophoresis, and diffusion*, Clarendon press, Oxford, England, 1974
6. Mächtle W.; *Progr. Colloid Polym. Sci.*, **1999**, 113, 1
7. Schmidt B., Riesner D. in *Analytical ultracentrifugation in biochemistry and polymer science*, Harding S.E., Rowe A.J., Horton J.C., Eds., The royal society of chemistry, Cambridge, England, 1992, 176
8. Cantow H.J., *Makromol. Chem.* **1964**, 70, 130
9. Scholtan W., Lange H., *Kolloid Z. Z. Polym.* **1972**, 250, 782
10. Müller H.G., *Colloid Polym. Sci.* **1989**, 267, 1113
11. Mächtle W. in *Analytical ultracentrifugation in biochemistry and polymer science*, Harding S.E., Rowe A.J., Horton J.C., Eds., The royal society of chemistry, Cambridge, England, 1992, 147
12. Schachman H.K; *Ultracentrifugation in Biochemistry*, Academic Press, New York, 1959
13. Cölfen H. in *Encyclopedia of Nanoscience and Nanotechnology* Nalwa H.S., Ed., American Scientific Publishers, 2003 in print
14. Stafford W.F. in *Analytical ultracentrifugation in biochemistry and polymer science*, Harding S.E., Rowe A.J., Horton J.C., Eds., The royal society of chemistry, Cambridge, England, 1992, 359
15. Schilling K.; Ph.D. thesis, Potsdam University, Germany, 1999
16. van Holde K.E., Weischet W.O., *Biopolymers* **1978**, 17, 1387
17. Demeler B., Saber H., Hansen J.C.; *Biophys. J.* **1997**, 72, 397
18. Geiselmann J., Yager T.D., Gill S.C., Camettes P., von Hippel P.H., *Biochemistry* **1992**, 31, 111
19. Gill S.C., Yager T.D., von Hippel P.H., *J. Mol. Biol.* **1991**, 220, 325
20. Hansen J.C., Lohr D., *J. Biol. Chem.* **1993**, 268, 5840
21. Hansen J.C., Ausio J., Stanik V.H., van Holde K.E., *Biochemistry* 1989, **28**, 9129
22. Schuck P., Perugini M.A., Gonzales N.R., Howlett G.J., Schubert D., *Biophys. J.* 2002, **82**, 1096
23. Yphantis D.A., *Biophys. J.* **1984**, 45, 324a
24. Stafford W.F., *Anal. Biochem.* **1992**, 203, 295
25. Stafford W.F. in *Modern analytical ultracentrifugation,*. Schuster T.M. and Laue T.M.; Birkhäuser, Boston, MA, Basel, Switzerland Berlin, Germany, 1994, 119
26. Behlke J., Ristau O., *Eur. Biophys. J.* **1997**, 25, 325
27. Behlke J., Knespel A., *J. Cryst. Growth* **1996**, 158, 388
28. Lamm O.; *Ark. Math. Astron. Fysik* **1929**, 21B, Nr. 2, 1
29. Holladay L.A., *Biophys. Chem.* **1979**, 10, 187
30. Behlke J., Ristau O., *Progr. Colloid Polym. Sci.* **1997**, 107, 27
31. Holladay L.A., *Biophys. Chem.* **1980**, 11, 303

32. Philo J. in *Modern analytical ultracentrifugation,*. Schuster T.M. and Laue T.M.; Birkhäuser, Boston, MA, Basel, Switzerland Berlin, Germany, 1994, 156

33. Behlke J., Ristau O., *Biophys. J.* **1997**, 72, 428

34. Philo J., *Biophys. J.* **1997**, 72, 435

35. Demeler B., Saber H., *Biophys. J.* **1998**, 74, 444

36. Schuck P., McPhee C.E., Howlett G.J., *Biophys. J.* **1998**, 74, 466

37. Schuck P., *Biophys. J.* **1998**, 75, 1503

38. Schuck P., Millar D.B., *Anal. Biochem.* **1998**, 259, 48

39. Schuck P., *Biophys. J.* **2000**, 78, 1606

40. Rinde H. Ph.D. thesis, University of Upsala, Sweden, 1928

41. Nichols J.B., *Physics* **1931**, 1, 254

42. Nichols J.B., Kramer E.O., Bailey E.D., *J. Phys. Chem.* **1932**, 36, 326

43. Cölfen H., Habilitation thesis, University of Potsdam, Germany, 2001

44. Cölfen H., Schnablegger H., Fischer A., Jentoft F.C., Weinberg G., Schlögl R.; *Langmuir* **2002**, 18, 3500

45. Cölfen H., Pauck T., *Colloid Polym. Sci.* **1997**, 275, 175

46. Cölfen H., Pauck T., Antonietti M., *Progr. Colloid Polym. Sci.* 197, **107**, 136

47. Rapoport D.H., Vogel W., Cölfen H., Schlögl R.; *J. Phys. Chem. B* **1997**, 101, 4175

48. Kurth D.G., Lehmann P., Volkmer D., Cölfen H., Koop M.J., Müller A., DuChesne A., *Chem. Eur. J.* **2000**, 6, 385

49. Börger L., Cölfen H., Antonietti M., *Colloid Surface A* **2000**, 163, 29

50. Börger L., Cölfen H., *Progr. Colloid Polym. Sci.* **1999**, 113, 23

Chapter 9

Recent Developments on Resolution and Applicability of Capillary Hydrodynamic Fractionation

J. Gabriel DosRamos

Matec Applied Sciences, 56 Hudson Street, Northborough, MA 01532

Abstract

Capillary Hydrodynamic Fractionation (CHDF) is a high-resolution particle size distribution (PSD) analysis technique. CHDF is used to measure the PSD of colloids in the particle size range of 5 nm to 3 microns.

CHDF fractionation occurs as an eluant or carrier fluid carries the particles downstream in a capillary tube. Large particles exit the fractionation capillary ahead of smaller particles. Particle fractionation occurs because of the combination of the eluant parabolic velocity profile (laminar flow), size exclusion of the particles at the capillary wall, and colloidal forces.

The aim of this study was to expand CHDF's applicability to a broader class of colloidal systems. This can widen CHDF's usefulness as a particle sizing technique.

Introduction

Particle sizing techniques can be grouped into High-Resolution (Fractionation) and Ensemble techniques *(1)*. High-Resolution (HR) techniques are characterized by the fact that particles are fractionated according to size and/or mass during particle size analysis. Ensemble techniques perform measurements on all particles simultaneously without physical separation. HR particle size analyzers include Capillary Hydrodynamic Fractionation (CHDF), Field-Flow Fractionation, Single-Particle Counting, and Disc Centrifugation. Ensemble techniques include Laser Diffraction, Photon-Correlation Spectroscopy (PCS), acoustic-attenuation spectroscopy, and Turbidimetry. Electron Microscopy is in a class by itself as it offers high resolution but particles are not fractionated during analysis.

HR techniques offer a strong advantage in that they produce *true* particle size distribution (PSD) data. HR-based devices can in principle detect the presence of multiple particle size populations without making significant assumptions. On the other hand, HR devices tend to be more complicated to operate than ensemble instruments. Ensemble instruments produce mainly mean particle size and standard deviation data. Any mean particle size value can be produced by an infinite number of PSD curves. This ill-conditioned problem, along with the fact that calculated PSD's vary significantly with minor noise in the raw data, force most ensemble devices to assume *a priori* the shape of the PSD *(2)*. Despite these issues, ensemble devices are much more widely used than HR instruments. The two main reasons seem to be the following. Ensemble instruments are easier to use, and are more widely applicable to different types of dispersion/colloidal samples.

Figure 1 describes the particle size-based fractionation process in CHDF. Larger particles exit the fractionation open capillary ahead of smaller ones *(3)*. A UV-detector is typically used as a particle-concentration detector. Such particle fractionation occurs because of a particle-size exclusion effect, plus colloidal forces affecting the particle motion. The latter consist mainly of particle/capillary electric double layer repulsion, and a lift force exerted by the moving fluid on the particles *(4)*.

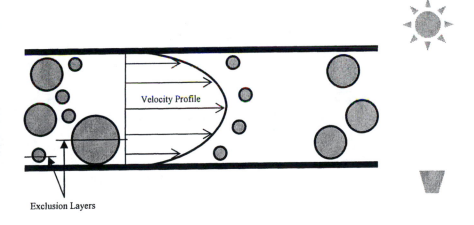

Figure 1. Particle size-based fractionation in CHDF.

Figure 2 shows CHDF2000 PSD data from two different polystyrene samples with the same volume-average mean particle size value of 226 nm even though their PSD's are noticeably different. This data exemplifies the risks of relying exclusively on mean particle size data. These samples will behave differently despite their identical mean particle size.

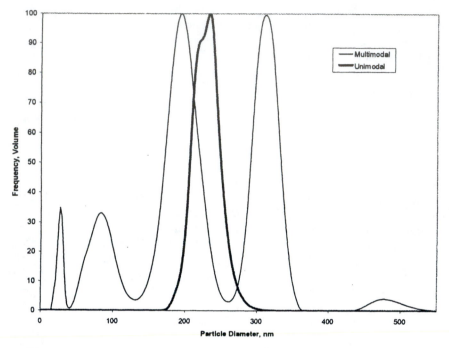

Figure 2. Two different samples analyzed by CHDF. Both have the same Volume-average (mean) particle size of 226 nm even though the PSD's are different.

It is in principle possible for a sample with a 226 nm mean particle size not to contain any 226 nm particles. This constitutes another disadvantage of relying on mean particle size data alone.

This paper describes efforts to expand the applicability of CHDF to different types of dispersed systems, including expanding its particle size analysis range. Also, a process on-line CHDF setup with automated sample dilution is presented.

Experimental

CHDF measurements were performed using a commercial CHDF2000 high-resolution particle size analyzer from Matec Applied Sciences, Northborough, MA *(5)*. Various colloidal samples were used. Polystyrene samples were obtained from Duke Scientific (Indianapolis, IN), and Seradyn (Indianapolis, IN).

Nanoparticle Size Analysis

Small particles, especially under 50 nm, are becoming more widely manufactured and employed in various intermediate and final colloidal products. Accurate particle size analysis of these small particles is essential. Small particles offer a large total surface area. Secondary smaller particle size populations can be unexpectedly present in any dispersion. Such smaller particles can sharply influence the performance of any dispersion. Additionally, a small number of larger particles can posse difficulties, e.g., larger particles in inkjet printing inks can plug ink conduits in today's inkjet printers.

Particle sizing of nanoparticles is difficult for most particle sizing techniques. Because of refractive index issues, as well as the fact that larger particles mask smaller particles, ensemble-type measurements such as Laser Diffraction and Photon Correlation Spectroscopy (PCS) have difficulty analyzing such nanoparticles. High-resolution devices such as disc centrifuges are also limited due to the lack of tendency of nanoparticles to sediment, even under strong centrifugal fields.

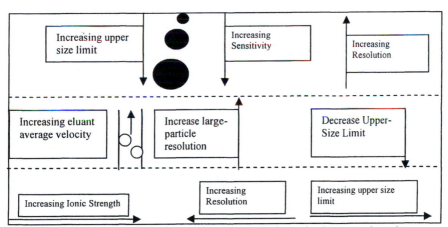

Figure 3. Effects of capillary inner diameter and eluant ionic strength and average velocity on CHDF resolution and particle size fractionation range.

Fig. 3 illustrates the effects of capillary ID, eluant average velocity, and eluant ionic strength on the resolution and particle-size fractionation range. As the capillary ID increases, so does the sample volume in the capillary. This increase in sample volume is due to the fact that the waste/fractionation split ratio changes *(6)*, i.e. more sample flows into the fractionation capillary relative to the "waste" stream. Such increase in fractionated-sample volume results in stronger particle detection in the CHDF particle-concentration detector, usually a UV detector.

Conversely, fractionation resolution increases as the capillary ID is reduced *(7)*. This is due to several compounding factors as follows: (i) the particle exclusion layer is larger relative to the capillary ID, (ii) the lift force is stronger, and (iii) the particle fractograms are narrower due to lower axial dispersion.

The upper particle size limit decreases with decreasing capillary ID. Physically, larger particles can flow in a larger-ID fractionation capillary; also, the lift force is lower for larger-ID capillaries; this allows fractionation among larger particles.

The eluant average velocity also plays a role on the resolution and particle size range. As the eluant average velocity in increased, large-particle size fractograms become narrower. This Fractogram narrowing is due to the increase in lift force strength which forces larger particles to travel closer together. As the lift force increases, the upper particle size fractionation limit decreases similarly to reducing the capillary ID *(8)*.

Suitable capillary ID and length, plus the eluant ionic strength and mean velocity can be combined to produce high resolution PSD data as shown in figures 4 and 5. Data is shown for the fractionation of eight particle size populations in less than 10 minutes.

This blend is composed of (from left to right on the raw-data graph) 800, 605, 420, 310, 240, 150, 60, and 20 nm. The peak separation on the raw data graph can be enhanced further by simply lengthening the fractionation time. The raw data peaks appear more overlapped than on the PSD. The reason is that a deconvolution procedure has been applied in the PSD computations *(9)*. The deconvolution computations are similar to those used in Gel Permeation Chromatography for incorporating into the PSD computations axial dispersion of particles during capillary flow. Axial dispersion broadens the Fractogram width.

CHDF offers a useful alternative for nanoparticle analysis due to the fact that larger particles do not mask the presence of small particles, as well as, nanoparticle analysis is as easy and accurate as for larger particles. However, CHDF also faces difficulties in analyzing particles smaller than 20 nm. These

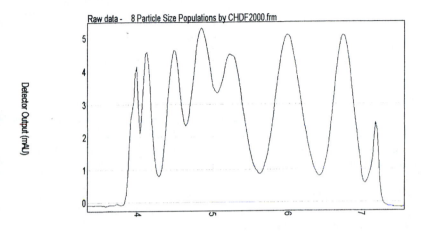

Figure 4. CHDF fractionation UV-detector raw data for a blend of 8 polystyrene calibration standards.

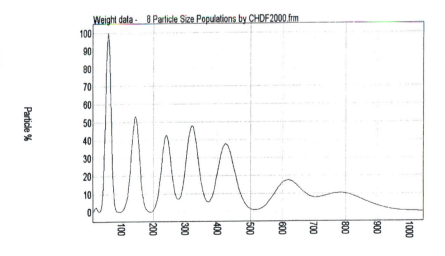

Figure 5. CHDF PSD data for an 8-mode polystyrene blend.

particles are difficult to quantify in the presence of larger particles due to large differences in UV-light extinction cross section *(10)*. These difficulties are shown in the well-known Beer-Lambert's law equation as follows:

$$D.O. = N \text{ Rext } \chi \qquad\qquad [1]$$

Where D.O. is the UV-detector output, N is the number of particles per unit volume, Rext is the particle extinction cross section, and χ is the UV-detector flow-cell path length. Equation [1] allows the calculation of N for each individual slice of particles exiting the CHDF fractionation capillary. N computations are susceptible to minor noise in D.O. when a sample contains particles under 20 nm along with larger particles, e.g. particles over 500 nm.

Figure 6 shows an extinction cross section curve for polystyrene particles in water. This curve was generated from Mie-theory computations built into the operating software on the commercial CHDF2000 device.

Figure 6 shows that there are several orders of magnitude between the Rext of particles smaller than 20 nm and those larger than 500 nm. Consequently, small D.O. errors become largely amplified in the computation of N.

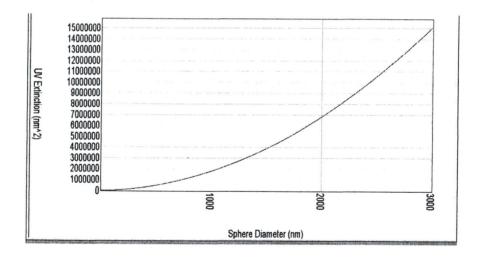

Figure 6. Extinction cross section for polystyrene particles in water. Curve generated from Mie theory computations.

Despite the challenges described above, the CHDF2000 device was able to accurately perform particle sizing measurements of silica particles under 20 nm as shown in Figures 7 and 8.

The accurate measurement of these small particles was achieved by maximizing resolution and particle-detection sensitivity. This required optimizing the combination of suitable capillary diameter and length, and eluant ionic strength and average velocity.

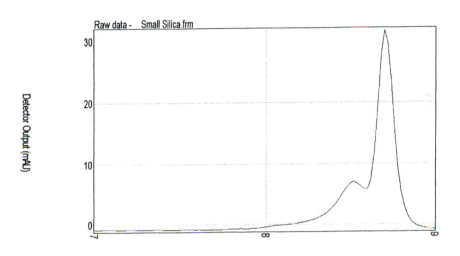

Figure 7. CHDF UV-detector raw data output for a 5-nm nominal particle size silica.

The CHDF raw data in figure 7 shows two main peaks located at 8.5 and 8.7 minutes. The peak at 8.7 minutes is believed to originate from UV-absorbing molecules in the sample such as surfactants, electrolytes, and acid or base molecules *(11)*. These molecules exit the capillary last because their particle size is smaller than that of the silica particles. The molecular peak does not appear in the PSD graph of figure 8 because its particle size is below the low computational limit of 1 nm.

Figure 8 presents the PSD for this silica sample. A lognormal PSD shape is obtained ranging from 1 nm to 30 nm. The mode is located at 4 nm.

In order to achieve the maximum resolution required for these small particles *(7)* a low ionic strength (0.1 mM) carrier fluid was used in conjunction with a 5-micron ID fused silica capillary.

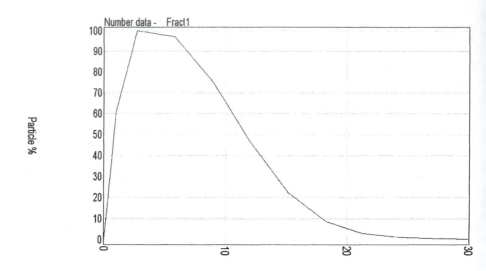

Particle Size (nm)

Figure 8. CHDF PSD for a 5-nm nominal particle size silica sample.

CHDF Particle Size Analysis of Micron-Sized Particles

CHDF has been commonly used for analysis of particles below 1 micron in size. As mentioned above, larger particles are typically subject to a "Lift Force" in the fractionation capillary. The lift force pushes larger particles toward the center of the capillary. The Lift force is proportional to the ratio of particle to capillary radii, and to the eluant average velocity. Thus, lowering the eluant average velocity along with using larger-ID fractionation capillaries reduce the lift force and extend the CHDF particle size upper limit.

Fig. 9 shows CHDF UV-detector vs. time fractionation data for a blend of 5, 1, and 0.1 micron polystyrene standards. The marker is injected about one minute after the blend. This data shows that CHDF fractionation is indeed able to fractionate particles larger than one micron. However, it is desirable to increase the fractionation resolution further in order to enhance particle size analysis capability for these larger particles. Further work is in progress.

On-Line Development

Today, there is a lack of (process) in-line, at-line, or on-line particle size analyzers suitable for analysis of liquid dispersions. Process particle sizers can become a vital component of slurry/dispersion/latex/emulsion production. Such analyzers can provide labor savings, as well as ensure product quality. Process particle sizers must be highly accurate, reproducible, precise, and reliable *(12)*.

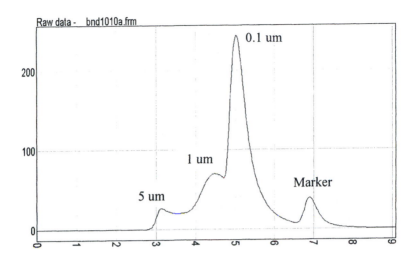

Raw data - bnd1010a.frm

Time (Min.)

Figure 9. CHDF UV-detector raw data output for a blend of 5, 1, and 0.1 micron polystyrene latex standards. Sodium benzoate "marker" is injected about 1 minute after the blend.

Fig. 10 shows a schematic diagram of an on-line CHDF device. The CHDF2000 unit used here is identical to the off-line device available commercially (Matec Applied Sciences, Northborough, MA). With the off-line device, samples are injected into the fractionation capillary using either an on-board manual HPLC-type injection valve or an HPLC auto-sampler. In the on-line setup, the sample flows directly from the process into an automated injection valve. The automated injection valve is actuated by the CHDF2000 on-line software in order to make sample injections every 5-10 minutes.

This particular arrangement has been used with a batch-polymerization reactor. This on-line device can also be used with continuous or semi-continuous reactors, as well as steady-state slurry streams.

The percent solids level of a batch polymerization reactor starts at zero and increases with time *(13)*. The on-line CHDF setup must be able to handle this constant increase (CHDF analysis is typically performed in the weight-percent solids range of 0.1 to 5%). The analysis process is as follows:

A "drip" line connected to the reactor system takes a steady stream of sample through the remotely-actuated (HPLC) injection valve. The sample is diluted as needed at point D. The injection valve automatically takes samples from the drip line and performs sample injections into the CHDF fractionation capillary every few minutes. CHDF eluant continuously flows through the injection valve in order to carry the samples into the fractionation capillary. The dilution valve sets the diluent vent (Y)/dilution (D) split ratio.

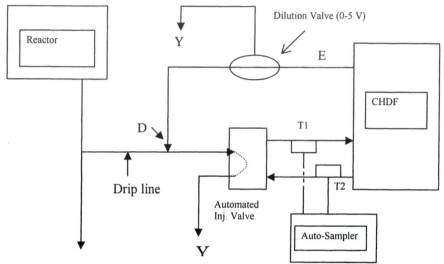

Figure 10. On-line CHDF setup equipped with sample auto-dilution.

In order to eliminate the need for a diluent pump, the effluent (E) from the CHDF is used as diluent. E is split at the dilution valve into a vented (Y), and a diluent (DFR) portion.

Valves T1 and T2 allow the use of an HPLC auto-sampler for calibration-standard analysis. The operator can thus perform periodic performance tests of the on-line CHDF setup.

The CHDF software calculates the area under each Fractogram. The dilution level is calculated by comparing the current Fractogram area to a pre-established suitable area value as follows:

$$DFR = C*PA/FA \qquad\qquad [2]$$

Where DFR is the diluent flow rate, C is a constant, PA is the pre-established (acceptable) Fractogram area, and FA is the sample fractogram area. DFR is set by sending a 0-5 volt signal to the dilution valve. As DFR increases, the vented (Y) eluant flow rate decreases.

PSD data files are saved to a network location. A Honeywell *PlantScape* process control system reads the PSD data and performs process control steps such as valve opening/closing and reactor temperature adjustment.

Figures 11 and 12 show polyvinyl acetate CHDF data collected from an on-line device connected to a batch reactor.

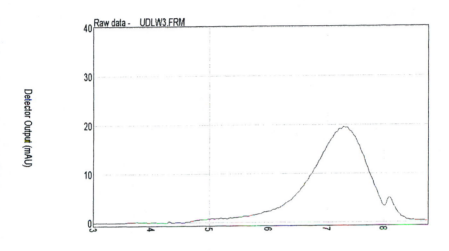

Figure 11. On-line CHDF UV-detector output for a PVA latex from a batch reactor.

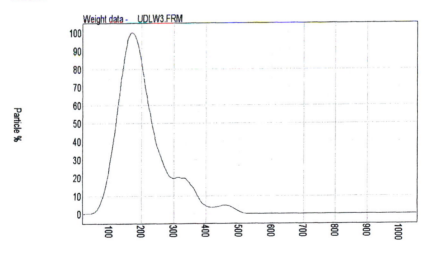

Figure 12. On-line CHDF PSD data for PVA latex from a batch reactor.

150

Conclusions

CHDF particle size fractionation can be used for high-resolution particle size analysis of dispersions in the particle size range of 2 nm to 5 microns. A process on-line particle sizer has been implemented based on CHDF fractionation. This on-line device is capable of performing automatic sample dilution, and interfacing with a Honeywell Plant Control system.

Acknowledgements

The author would like to thank Dr. Tim Crowley for sharing some of the CHDF on-line data presented here.

References

1. Barth, H. G., and Flippen, R. B., *Anal. Chem.*, 67, 257R-272R, **1995**.

2. Weiner, B. B., and Tscharnuter, W. W., in *Particle Size Distribution: Assessment and Characterization*, ACS Symp. Series 332, p. 48, 1987.

3. Silebi, C. A., and DosRamos, J. G., *AIChE J.*, 35, 165, 1989.

4. DosRamos, J. G., Ph.D. Dissertation, Lehigh U., 1988.

5. www.matec.com

6. DosRamos, J. G., and Silebi, C. A., *J. Coll. Int. Sci.*, 135, 1, 1990.

7. Venkatesan, J., DosRamos, J. G., and Silebi, C. A., in *Particle Size Distribution II: Assessment and Characterization*, ACS Symp. Series 472, p. 279, 1991.

8. DosRamos, J. G., in *Particle Size Distribution III: Assessment and Characterization*, ACS Symp. Series 693, p. 207, 1998.

9. Silebi, C. A., Ph.D. Dissertation, Lehigh U., 1977.

10. Bohren, C., and Huffman, D. R., *Absorption and Scattering of Light by Small Particles,* Wiley Interscience Publication, 1983.

11. DosRamos, J. G., and Silebi, C. A., *Polym. Int.,* 30, 445, 1993.

12. Venkatesan, J., and Silebi, C. A., in *Particle Size Distribution III: Assessment and Characterization*, ACS Symp. Series 693, p. 266, 1998.

13. Dr. Tim Crowley, U. Delaware, direct communication.

Chapter 10

Capillary Hydrodynamic Fractionation of Organic Nanopigment Dispersions

John Texter

College of Technology, Eastern Michigan University, 122 Sill Hall, Ypsilanti, MI 48197

Capillary hydrodynamic fractionation (CHDF) is generally useful for sizing particles in the range of 30-800 nm in equivalent spherical diameter. This fractionation method uses capillaries of 10-15 μm inside diameters. Such hydrodynamic fractionation of particles according to their equivalent diameter comes about because of Poiseuille flow and the tendency of larger particles to sample higher velocity components of the flow field. This paper describes the application of CHDF to a series of organic nanopigment dispersions used in photographic films and papers to manage light transmission and reflection. The results are benchmarked against analyses of the same dispersions by sedimentation field flow fractionation and by disk centrifugation. Transmission electron micrographs are presented to allow for an independent check of the results. The CHDF and SFFF results are in good agreement and may be viewed as competitive techniques. CHDF measurements are much faster, and SFFF measurements allow for finer resolution.

Introduction

Organic nanopigments are becoming increasingly important because of photographic (*1,2*), printing (*3,4*), inkjet (*5-7*), and other marking and display technologies. Organic nanoparticles are also becoming extremely important in drug delivery and targeting (*8,9*). Such particulates tend to be made by comminution processing (*10-12*) or by various forms of condensation (*13-16*). There are only a few reported examples where organic nanoparticles are being directly precipitated (*17*), but such reports are likely to grow as volume demand for such nanoparticles increases. Whatever the ultimate application, means are needed to quantify particle size and polydispersity, and in this chapter we focus on the applicability of capillary hydrodynamic fractionation (CHDF) to sizing organic nanopigments that have been demonstrated to be useful in a wide variety of photographic products (*18*).

While CHDF is the focal point of this work, we examine it from a benchmarking perspective in comparison to two competitive techniques, sedimentation field flow fractionation (SFFF) and disk centrifugation (DC).

Elements of CHDF

CHDF has largely supplanted hydrodynamic chromatography(*19-20*) as an analytical technique. Hydrodynamic chromatography utilizes a column filled with monodisperse beads, for example, in the size range of 10-250 μm. The beads are impervious to the sample solvent and small particulates suspended therein. The multiphase flow is driven through the column in many parallel paths of varying tortuosity, but each path at an instant may be envisioned as a flow through an interstice between three beads in axisymmetric trigonal contact with one another. The flow through the "center" along the symmetry axis will have the greatest velocity component and the flow creeping along the bead surfaces will have the slowest velocity component. Smaller particles may approach the beads more closely than larger particles, and on average will sample a slower hydrodynamic flow field. This is the basis of hydrodynamic chromatography. Larger particles pass through more quickly than smaller particles, and are thereby separated or fractionated. This process is shown schematically in Figure 1. Better separation is obtained with smaller diameter columns. A marker is injected, and retention time calibrated against the marker elution. Particle transport depends on the size of the packing particles, ionic strength, flow velocity, and eluting particle size. Naturally, agglomerates and flocs, when present, will be analyzed as if they are single particles. Thus stability of dispersions may be studied by comparing particle size distributions obtained as a function of time or treatment.

CHDF is a single flow-path analogue of hydrodynamic chromatography and was invented by Silebi and Dos Ramos (*22*). The essential element of CHDF is the hydrodynamic laminar flow of a particle through a very small capillary. Fused silica capillaries with internal diameters as small as 4 μm are readily commercially available. Hagan-Poiseuille flow in such a microcapillary is

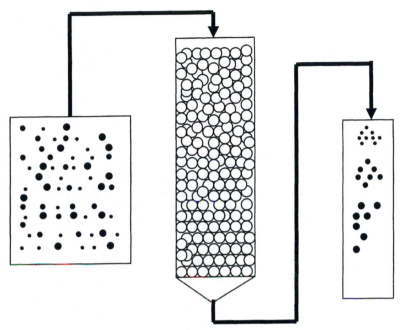

Figure 1. Separation of suspended particles according to effective size by hydrodynamic chromatography (*21*). © 2000 Strider Research Corporation.

shown schematically in Figure 2, for flow from the axisymmetric axis at r = 0 to the capillary wall. The parabolic fluid velocity components are qualitatively illustrated and show how the magnitude of the fluid velocity changes from a minimum at the walls of the capillary, to a maximum along the symmetry axis of the capillary. This fluid velocity profile is given quantitatively by the Hagan-Poiseuille equation:

$$d^2 = 32\eta L \frac{\langle v \rangle}{\Delta P}$$

where d *is* the capillary inner diameter in meters, η is the viscosity of the fluid in Pascal seconds, L is the capillary length in meters, $<v>$ is the average velocity of the fluid in meters per second, and ΔP is the pressure drop across the tube (inlet/outlet) in Pascals. It is known that the parabolic velocity profile depicted in Figure 2 yields an average $<v>$ velocity that is half the maximum velocity, v_{max}, seen in the center of the capillary:

$$\langle v \rangle = \frac{1}{2} v_{max}$$

154

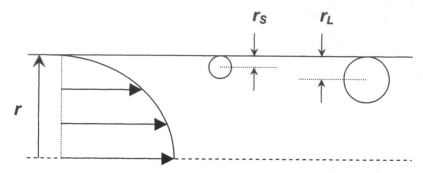

Figure 2. Hagan-Poiseuille flow in a microcapillary; the physical basis of capillary hydrodynamic fractionation. The flow field is illustrated from the axisymmetric capillary axis (r=0) to the inner capillary wall. Large (r_L) and small (r_S) particles are illustrated.

As illustrated in Figure 2, larger particles can approach the capillary wall only to within a distance corresponding to their radius, r_L, while smaller particles can approach to within distances $r_S < r_L$. Smaller particles, therefore, sample, on average, a more slowly moving flow field, and larger particles are thereby hydrodynamically fractionated from smaller particles. The dynamic range of hydrodynamic velocities is 2. At the end of the capillary a microflow cell us used to monitor turbidity at a chosen wavelength. This optical density is used to estimate the particle number density for a given particle size. Latex standards are typically used to calibrate particle size as a function of elution time. These two aspects are combined to produce number frequency and volume frequency particle size distributions, along with various desired moments, such as number mean and volume mean diameters. This turbidimetric or optical density method of estimating particle number densities is perhaps one of the most severe assumptions implicit in the technique. While optical density in the UV (e.g., at 220 nm) is typically used, one may record densities at any wavelength accessible in the instruments UV-vis spectrophotometric detection system. At longer wavelengths, particularly in the visible, particle size effects, such as scattering in addition to absorption, will come into play. The highest accuracy will require the number density deconvolution to be done using optical constants appropriate for the specific materials being studied. Such procedures are detailed in the articles listed below.

The reader is referred to excellent theses by Dos Ramos (*23*) and by Venkatesan (*24*), and to various articles (*25-35*) for detailed discussions of the underlying principles of CHDF, along with discussions of particle resolution, effects of eluant solution composition on size discrimination, etc.

Elements of SFFF

SFFF bears a lot of similarity to CHDF, in that particles are pumped through a flow field. However, the particles are also fractionated by a centrifugal field, and it is the smallest particles that elute first. A diagram similar to that of Figure 2 is illustrated in Figure 3 to illustrate the essential aspects. Particles are forced through the SFFF channel, and flow is again parabolic as given by the Hagan-Poiseuille equation above. One half of the essentially symmetric flow field is illustrated from the symmetry axis at $r=0$ to the outer channel wall. A perpendicular centrifugal force, F_c, drives particle sedimentation toward the outer wall. Larger particles sediment more quickly in this centrifugal field than do smaller particles. In this case smaller particles sample, on average, a higher flow velocity field and are thereby fractionated from larger particles. Smaller particles, therefore, elute ahead of larger particles (the reverse of the order obtained in CHDF). The dynamic range of SFFF is in principle larger than that of CHDF, because the effects of the centrifugal force field can be dramatic. However, this benefit comes at the expense of much longer measurement times, since a factor of ten decrease in particle size requires a factor of 100 increase in sedimentation time to reach the same radial displacement. Particle detection is done by measuring optical turbidities. Other elements and applications are discussed in detail in Allen's monograph (*36*).

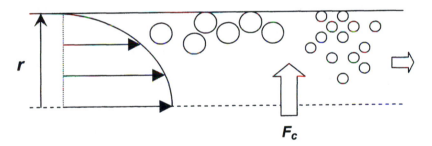

Figure 3. Forces and flows in sedimentation field flow fractionation (SFFF). Flow and fractionation occur parallel to the channel symmetry axis (dotted line), and a laminar Hagan-Poiseuille flow field hydrodynamically drives the elution process. This flow occurs in channels in a centrifuge rotor, and a perpendicular centrifugal force, F_c, drives particle sedimentation toward the outer wall. Larger particles sediment more quickly in this centrifugal field than do smaller particles.

Elements of Disk Centrifugation

The line-start method of DC involves layering an aliquot of test dispersion upon a more dense sedimentation or spin fluid, centrifuging the rotor to cause the particles to centrifuge down through the spin fluid, and then measuring the turbidity as particles pass an optical detector as a function of the spin time. Larger particles sediment faster, so the sample is fractionated during sedimentation. A calibration curve is used to convert spin times to equivalent diameter, and another calibration curve or mathematical algorithm is used to convert detector turbidities to particle number densities. In principle, DC has a larger dynamic range than SFFF, and much larger than CHDF, because of the inverse square relationship between particle diameter and sedimentation time to an observation radius, r. This relationship is given by the equation:

$$t = \frac{18\eta \, \ln(r/r_1)}{\Delta\rho\omega^2 d^2}$$

where t is the time for a particle to sediment from a starting radius r_1 to an observation radius r, η is the spin fluid viscosity, $\Delta\rho$ is the particle-spin fluid density difference, d is particle diameter, and ω is the radial velocity.

As with CHDF and SFFF, DC also uses turbidity detection to estimate particle number densities, and hence suffers similarly from the assumptions and limitations of such estimations. The precise nature of corrections required to correct for these assumptions varies with the material being analyzed and the wavelength at which detection of turbidity is made. Discussions of the technique are given by Oppenheimer (37), Coll and Searles (38), and Allen (39).

Most particle size measuring techniques, including those examined here, treat isolated and separate particles and aggregates and tight flocs similarly, as long as the aggregate moves as a single body. These methods, therefore, do not provide direct information on the structure of particles. Microscopy methods are very helpful for discerning aggregation effects.

Objectives

Our benchmarking study summarized in this chapter focuses on organic nanopigments that are generally fairly polydisperse, and range in size from 20 nm to over 800 nm in largest dimension. Aliquots of dispersion of these nanopigments were analyzed by three ensemble techniques that rely on particle fractionation according to size. We also compare these results with more visceral data obtained by transmission electron microscopy, as a reality check. We have not pursued TEM-based image analysis to generate particle size distributions, although one of the pigments reported on here (**Y1**) has been examined by detailed image analysis (1). A preliminary report of this study has been presented earlier (40).

Materials and Methods

Pigments

The organic pigments **C1**, **M1**, **M2**, **Y1**, **Y2**, and **Y3** are illustrated in Table 1 and were obtained from Eastman Kodak Research Laboratories (Rochester, NY). Dispersions of these pigments were prepared by small media milling techniques similar to those used in manufacturing these nanopigment dispersions for various photographic microfilm, x-ray film, graphic arts film, and color reversal film products and applications. The various nanopigment dispersions were prepared at pigment weight fractions of 4-20% in water and all were stabilized by anionic surfactants, either OMT (oleoylmethyltaurine, sodium salt) or TX-200. The properties of these dispersions are tabulated in Table 2.

SFFF Analyses

The dispersions were diluted about 4-fold with carrier fluid. Aqueous oleoylmethyltaurine (sodium salt; OMT; 0.1% w/w) at pH 5.9 was used as carrier fluid and eluant, since the standard (alkaline) eluant dissolved the aqueous nanopigment particles. A DuPont Instruments SF3 1000 Particle Fractionator was used for the measurements and was operated at about 12,500 rpm to generate the fractionation and size distributions.

Disk Centrifugation

Analyses by disk centrifugation were diluted about 2-fold with pH 5.0 ± 0.1 water, and then diluted another 15-fold to 80-fold before injection. Samples were injected into a Joyce Loebl disk centrifuge. The particular dilution utilized was selected in order to obtain an appropriate level of turbidity in the optical detection system. The spin fluid was prepared by adding about 10 mL of 4% sucrose, followed by 1 mL of citrate acid buffer (4.61 g/L). The disk centrifuge was operated at 8,000 rpm. Samples were injected after spinning for about one minute.

Transmission Electron Microscopy

TEM analyses were done on dispersion samples that were diluted and sonicated for 1 minute on a Branson Sonifier 250 Probe Sonicator (output control setting #3). The samples were then spotted onto carbon coated copper grids in a low vacuum bell jar and vacuum dried. Pt/Pd shadowing was then done at an oblique angle of about 18°, and then coated with a very thin layer of carbon.

Table 1. Pigment Structures

Pigment	Structure
C1	
M1	
M2	
Y1	
Y2	
Y3	

Slurry samples containing gelatin were diluted and sonicated and then air-fuged onto carbon-coated grids at 40,000 rpm using a Beckman Airfuge air-driven ultracentrifuge. These samples were then shadowed and coated as described for the other slurries. A JEOL 100 CX11 transmission electron microscope was used to image these carbon replicas.

Table 2. Nanopigment Dispersion Formulations

Pigment	% Pigment (w/w)	Dispersing Aid
C1	4	OMT
M1	4	OMT
M2	15	OMT
Y1	15	OMT
Y2	15	OMT
Y3	15	OMT

CHDF

A Matec Applied Science CHDF 1100 (Hopkinton, MA) particle sizing system, was used with CHDF 1100 operating software for control of the HPLC pump turbidity detection system, and a digitally monitored Rheodyne injection valve (5 μL). Turbidity was measured at 220 nm, and a series of polystyrene latexes (Duke Scientific) was used to calibrate sizes. A pair of HPLC single-piston drives were ganged in series to drive the eluant flow through a tee junction type flow splitter, to create flows 1 and 2 (described in Figure 4). The CHDF column had a nominal inner diameter of about 11-15 μm. A sodium benzoate marker solution (0.25% w/w) was injected 45-120 seconds after sample injection, so as to provide a direct measurement of the average fluid velocity, $<v>$. This retention time was then used to convert the experimental fractogram to an absorbance vs. equilibrium diameter plot. The eluant solution used was Matec GR500 diluted 1:10 v/v with distilled water. Samples for injection were diluted with this eluant so that the injected suspension was about 1% (w/w) in pigment. After dilution the samples were shaken on a vortex type mixer for at least 15 seconds, and then quickly injected.

An operational CHDF system is illustrated in the schematic of Figure 4. The surfactant reservoir (6) contains a surfactant solution that serves as a carrier fluid and eluant. It is formulated to passivate the silica capillary wall against particle adsorption. This eluant is driven by high pressure pumps, operating in the neighborhood of 5,000 psi. Two significant flow channels are indicated: first, a flow (1) terminating in a tee (7) just prior to the UV detector (8); and second, a flow through the injection valve (3) and through the CHDF column (2). A duplex pump (5) using eluant (6) drives these flows. Data are tabulated, analyzed, and displayed using an interfaced computer (9). Prior to flow (1)

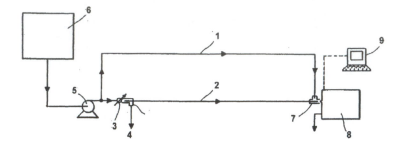

Figure 4. Block diagram of CHDF 1100 system (22). See text for operational discussion.

through the capillary, a flow splitter is configured in the injection valve (**3**). Dispersion samples are injected (5-25 µL) in front of the capillary, and a small volume fraction of the sample flows into the capillary to establish flow (**2**) through the CHDF capillary. This small volume element is fractionated hydrodynamically during the elution through the CHDF column. The bulk of this second flow passes around the capillary and out to waste (**4**). After the sample is fractionated in the CHDF column (**2**), it is diluted with make-up fluid coming from flow (**1**) at the tee junction (**7**) prior to the UV detector. This combined flow passes through the UV detector comprising a microflow cell, and optical turbidity is measured in absorbance units. This flow passes out of the detector and on to the waste stream. The optical turbidity (absorbance) measured as a function of retention time, or time after injection, is recorded and displayed as the experimental "fractogram".

A typical "fractogram" is illustrated in Figure 5 for a dispersion of nanopigment **C1** (see Table 1 for structure and Table 2 for formulation). The peak that is labeled "marker" indicates the optical absorption due to the (0.25%) sodium benzoate; the residence time of this marker serves to establish the average fluid velocity through the column, at a given set of operating parameters. The marker was typically injected about 60 seconds after sample injection. The operating software establishes the location (elution time) of this marker peak, and uses this time to establish the minimum detection time, labeled "min" in Figure 5, that corresponds to elution of the particles too small to be fractionated and UV-absorbing components in solution (typically dispersant or surfactant). The time labeled "max" in Figure 5 corresponds to the fastest elution time possible, and the elution of the largest possible particles. The experimental dynamic range of detection, in terms of elution time, is simply the ratio.

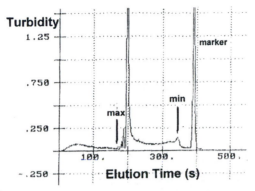

Figure 5. Experimental CHDF fractogram for the nanopigment **C1** dispersion.

illustrated of max to min, 1:2. The data in Figure 5 show a case where turbidity is detected over the entire dynamic range of the experiment with a small peak at about 350 s retention, corresponding to very, very small particles, and a very large peak at about 200 s retention corresponding to very large particles. These data will be discussed in much greater detail in the next section.

The raw data of Figure 5 were converted to number and weight (actually volume) frequency particle size distributions by numerically deconvoluting the fractograms. The ratio of the marker time to the sample elution time is called the selectivity ratio. Nominally monodisperse polystyrene latex particles ranging from 30-800 nm in diameter were used to calibrate the deconvolution software, to enable the generation of size distributions. Our procedures utilized eight such polystyrene "standards", about half were selected with size greater than 300 nm and half were selected with sizes below 300 nm. These calibrations were done separately for each individual capillary used. These calibration samples were presumed to be monodisperse, and their axial dispersion was measured and saved in memory, along with the elution time. The calibration diameters and selectivity ratios were used to construct a calibration curve of size as a function of selectivity, and the experimental axial dispersion was used to estimate sample axial dispersion as a function of diameter. This calibration establishes the correspondence between elution time and polystyrene-equivalent diameter. The turbidity measured at each elution time, corresponding to a particular polystyrene-equivalent size, was then converted to a particle number density corresponding to that same size, and detailed in the Dos Ramos thesis (23). A number density distribution function in terms of polystyrene equivalent spherical diameters was thereby produced. Likewise, a volume frequency distribution is constructed from this number distribution, and cumulative distributions of each are computed. Examples of these distributions are illustrated in Figure 7 and later figures as discussed in the next section.

Results and Discussion

Benchmarking

CHDF, SFFF, and DC results obtained for these six nanopigment dispersions are tabulated in Table 3. Number (d_N) and volume (d_V) mean diameters for CHDF and SFFF results show fairly good agreement and self-consistency. The CHDF and SFFF results have a higher degree of self-consistency with each other than between the CHDF and DC or SFFF and DC results.

The biggest discrepancy between the CHDF and SFFF results is for the case of the **C1** dispersion. This disparity is discussed at length in the next section. Otherwise, both number and volume means are in nearly quantitative agreement for four of the dispersions. The polydispersity as reflected in the quotients d_V/d_N tend systematically to be a little higher in the SFFF results (1.8 to 2.4, except 1.4 for **C1**), compared to the CHDF value (1.4 to 1.7, except 7.0 for **C1**), but most of these ratios are within 10-20% of each other. These d_V/d_N ratios for the DC results range from 1.1 to 1.4 (except 1.7 for **C1**) and are systematically much

Table 3. Comparison of Average Effective Diameters (d_N, number mean; d_V, volume mean) Obtained by CHDF, SFFF, and DC Methods.

Pigment	CHDF		SFFF		DC	
	d_N	d_V	d_N	d_V	d_N	d_V
C1	75±11	527±105	103	147	66±2	110±2
M1	59±1	94±5	35	85	62±2	72
M2	133±9	226±9	137	267	125±2	170±2
Y1	64±2	103±2	50	100	82±2	97±2
Y2	70±4	110±4	69	123	79±2	98±2
Y3	91±7	124±5	56	128	119±2	130±2

lower than the CHDF or SFFF results. The DC results are, however, always in the same range as the results from the other techniques, and we should mention that the DC estimates were made by approximating the distributions as lognormal.

Since the CHDF and SFFF results have the greatest pair-wise self-consistency, we can conclude that they are competitive from an accuracy perspective. Each method has limitations and weaknesses. A limitation of CHDF is its somewhat limited resolving power, and this is a simple result of the very short dynamic range in the hydrodynamic fractionation. SFFF has greater resolving power, but measurements take of the order of an hour per sample, as opposed to ten minutes per sample in CHDF.

Polydispersity and Comparison with TEM

It turns out, as illustrated below, that all of these nanopigments have nonspherical morphologies; they all derive from crystalline structures, so it is not surprising that they present as small crystallites, rather than as spheres. The CHDF, SFFF, and DC sizing techniques all yield final results in terms of equivalent-spherical diameters. Rod-shaped particles in a flow prefer to align themselves parallel to the flow direction. However, such alignment is unstable, and such particles will spontaneously tumble, and thereby behave as if they had some degree of spherical character, at least with respect to their interactions with channel walls and flow fields. Our objective in this section is to compare the CHDF results with transmission electron micrographs of each of the nanopigment dispersions, to see qualitatively whether the polydispersity presented in such TEMs is consistent with the number and volume frequency distributions.

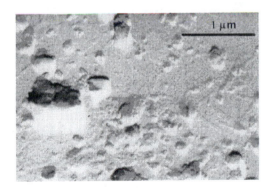

Figure 6. TEM of **C1** dispersion (carbon replicas were prepared and then shadowed with Pt/Pd.

The **C1** dispersion is particularly pathological. The fractogram in Figure 5 shows that this dispersion has particles spanning the entire dynamic range of the technique, from less than 30 nm to more than 800 nm in largest dimension. The results for **C1** in Table 3 indicate a number mean equivalent diameter of 75 nm and a volume mean diameter of more than 500 nm ($d_V/d_N = 7.0$). These data and the broad turbidity reported in Figure 5 indicate something we do not see very often. Most of the particles are less than 100 nm in diameter, but most of the mass appears to occur in much larger particles. A TEM of this **C1** dispersion is illustrated in Figure 6, and this TEM corroborates such a pathologically broad particle size distribution. This micrograph depicts a single large pigment particle nearly 800 nm long. From the shadow length we estimate the particle thickness to be about 300 nm. In the other extreme, there are very many very small particles evident with greatest dimensions less than 60 nm. Nearly the entire background of the TEM appears to be composed of such small particles.

These observations are supported quantitatively by the corresponding frequency and cumulative size distributions illustrated in Figure 7. The number frequency distribution shows unequivocally that most of the particles are very small, and the number average equivalent diameter is about 75 nm. However, the volume frequency distribution shows that the main mode is at about 700 nm, ten times larger than the number mode. The corresponding cumulative number distribution indicates that 90% of the particles are less than 100 nm in

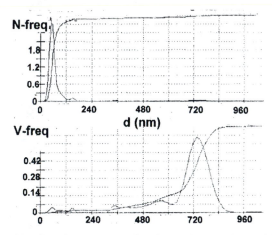

Figure7. Number frequency and volume frequency particles size distributions derived from CHDF data for the **C1** dispersion. Each of these frequency distributions is normalized. Also illustrated are the corresponding cumulative distributions, rising from 0% to 100% as equivalent spherical diameter increases from left to right.

size, while the cumulative volume distribution indicates that 90% of the sample nanopigment volume (or mass) is in particles larger than 400 nm. The large peak eluting at about 200 seconds in Figure 5 corresponds to this large size mode visible in the 600-800 nm region of the volume frequency distribution. The sharp and noisy peaks preceding this intense peak in the fractogram are typical of agglomerated particles. This sample also comprises many, many very small particles, but most of the mass is dispersed as relatively large particles. It is not exactly clear how or why this large mode fraction was missed in the SFFF experiments. The size difference between these two modes is so large, however, that the SFFF experiments likely was just too short in duration to detect the large particles.

Fractograms obtained for the other 5 nanopigment dispersions are illustrated in Figure 8. Each fractogram illustrates a marker peak on the far right and a steadily rising background turbidity. The turbidity peaks rising above these background slopes represent turbidity from the fractionated dispersions. Each of these fractograms presents a single turbidity peak, except for the case of **Y3**, which is an unequivocal doublet with a small peak at about 360 s and a major peak at about 315 s retention. Such a doublet is sufficient to show the existence of a distinctly bimodal size distribution. We will examine this aspect in detail later in this discussion.

M1

It is also apparent that the turbidity peak in the fractogram for **M1** is an unresolved doublet, with a peak at 310 s and a distinct shoulder at about 275 s. The corresponding number frequency and volume frequency distributions and a TEM of this dispersion are illustrated in Figure 9. The existence of a bimodal distribution is unequivocally seen in the volume frequency distribution in Figure 9(b) bottom. Modes at 60 nm and at about 140-180 nm are indicated. Most of the mass appears to be concentrated in the smaller mode. Both of these modes are clearly seen in the TEM of Figure 9(a). The distributions indicate 90% of the particles are smaller than 70 nm and about 53% of the particle mass (volume) is less than 70 nm. The larger mode accounts for the remaining 30% of the mass, but only 10% of the particles.

M2

A similar analysis may be made of pigment **M2**. The corresponding TEM and particle size distributions are given in Figure 10. The number frequency distribution tails off exponentially to higher diameters, indicating a broad distribution. This broadness is confirmed in the volume frequency distribution that appears to exhibit 6 modes, but suffice it to say the mass is spread fairly uniformly over particles in the range of 110-400 nm in equivalent spherical

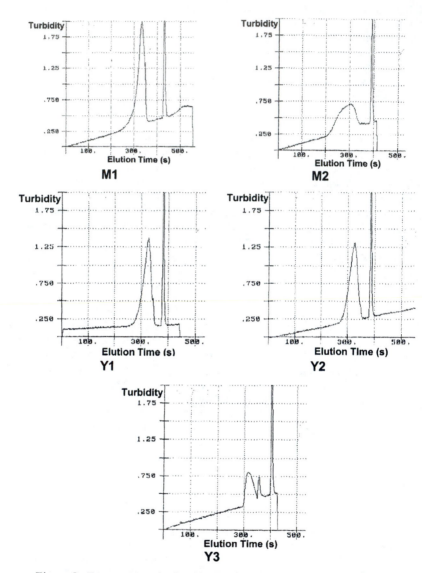

Figure 8. Fractograms obtained in CHDF measurements of **M1**, **M2**, **Y1**, **Y2**, and **Y3** nanopigment dispersions. Turbidity is displayed in units of optical density. The righthand most peak in each frame is a sodium benzoate marker peak.

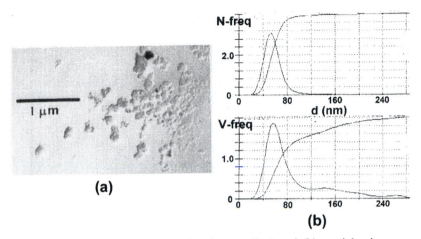

(a)

(b)

Figure 9. (a) TEM (shadowed carbon replica) and (b) particle size distributions for the dispersion of nanopigment **M1**. The corresponding cumulative distributions in (b) rise from 0% to 100% as equivalent spherical diameter increases from left to right.

diameter. The corresponding fractogram in Figure 8 exhibits a single turbidity peak that is fairly broad and rises from about 240 s to 360 s, peaking at about 330 s. The TEM in Figure 10(a) does not suggest a particular mode, but does indicate particles ranging from 100 nm to 500 nm in largest dimension are present, without any clearly resolvable number weighting.

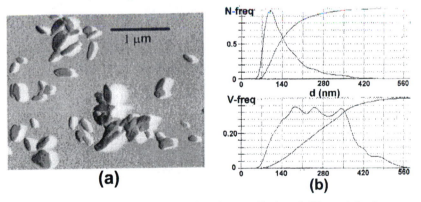

(a)

(b)

Figure 10. (a) TEM (shadowed carbon replica) and (b) particle size distributions for the dispersion of nanopigment **M2**.

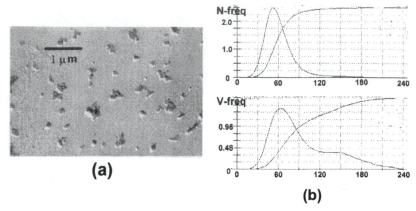

Figure 11. (a) TEM (shadowed carbon replica) and (b) particle size distributions for the dispersion of nanopigment **Y1**.

Y1

The fractogram for **Y1** in Figure 8 also exhibits an almost symmetric turbidity peak over the 260-360 s range. The size distributions in Figure 11(b) suggest that 60% of the particle mass is in the smaller mode peaking at about 60 nm, and that 30% of the mass is in a larger mode centered around 150 nm. The TEM of Figure 11(a) is consistent with these size distribution results, in that particles in both of these modes are clearly visible, but it is difficult to make any kind of qualitative weighting judgment from this micrograph.

Y2

The fractogram for **Y2** in Figure 8 also exhibits an almost symmetric turbidity peak over the. 275-360 s range, and looks very similar to that for **Y1**. The size distributions in Figure 12(b) appear more narrow than those for **Y1** in Figure 11(b). This is consistent with a 15% more narrow turbidity peak in the fractogram in Figure 8. About 63% of the mass is connected with the smaller mode associated with the peak at 65 nm, and about 30% of the mass is associated with a larger mode centered around a shoulder at about 150 nm. In the TEM of Figure 12(a) one can clearly see particles belong to these two modes. The volume frequency distribution indicates there is essentially no significant mass in particles greater than 220 nm in size.

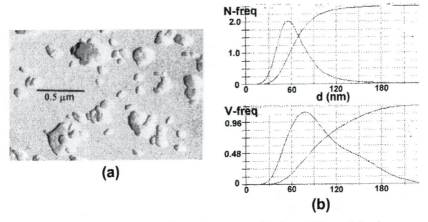

Figure 12. (a) TEM (shadowed carbon replica) and (b) particle size distributions for the dispersion of nanopigment **Y2**.

Y3

The **Y3** dispersion exhibits in Figure 8 a bimodal turbidity fractogram over the 300-360 s interval, and this bimodality is evident in both the number frequency and volume frequency distributions in Figure 13(b). The turbidity peak at about

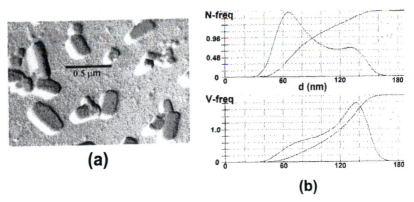

Figure 13. (a) TEM (shadowed carbon replica) and (b) particle size distributions for the dispersion of nanopigment **Y3**.

360 s in Figure 8 unequivocally indicates there are a lot of very small particles. The larger and broader peak at 315 s corresponds to the larger mass mode at about 138 nm. From the cumulative distributions in Figure 13(b), we see that about half the total number of particles are less than 80 nm, but these particles only account for about 14% of the total **Y3** nanopigment mass. The TEM in Figure 13(a) provides evidence for the very small particles and for the larger mode. The larger particles in Figure 13(a) have greatest dimensions in excess of the upper limits indicated in the volume frequency distribution of Figure 13(b). Analysis of the shadowing indicates, however, that these particles have widths of the order of 150 nm and thicknesses of the order of only 40 nm.

Conclusions

CHDF appears to very satisfactorily handle sizing nanopigment dispersions having equivalent particle diameters in the range of 30-800 nm. The comparison with SFFF suggests CHDF and SFFF are very competitive with respect to size distributions, although it should be stressed the CHDF measurements are much, much faster, and an independent measure of particle density is not required in CHDF (but is required in SFFF and disk centrifugation). These data strongly support the necessity of obtaining both number and volume frequency size distributions, in order to better understand the nature of polydispersity. As a practical matter, gelation and other untoward dispersion phenomena may arise from network formation among a large number of very small particles. It is known that such effects can occur for dispersions of **C1**, and the very high number of very small particles provides an explanation for such behavior, while a size measurement focusing only on volume or mass distribution would entirely miss such small particles. Applications of CHDF to ever smaller nanoparticles, such as those in the 10-50 nm range, should be feasible in smaller capillaries as well (*41*). All three techniques suffer from similar limitations in the variation of optical properties, even in the UV, with wavelength and with particle size. It generally appears that these limitations are more severe in the visible region or in non-absorbing regions, where the turbidity is entirely due to scattering. In this respect, turbidity detection with disk centrifugation typically is at a disadvantage relative to CHDF and SFFF where UV turbidity is typically used.

Acknowledgments

Thankful acknowledgment is made to Brian Morrin for making the SFFF measurements and to Mary Beth Mattern for making the disk centrifuge measurements. The transmission electron micrographs were kindly provided by Ken Schlafer and John Minter. Special thanks are extended to Eastman Kodak

Company for permission to publish this study and for the generous gift of a Matec Applied Sciences capillary hydrodynamic fractionation instrument.

Literature Cited

1. Texter, J., Electroacoustic characterization of electrokinetics in concentrated pigment dispersions: 3-Methyl-1-(4-carboxyphenyl)-2-pyrazolin-5-one monomethine oxonol, *Langmuir* **1992**, *8*, 291–298.
2. Brick, M.C.; Smith, T.M.; Factor, R.E.; Armour, E.A.; Bowman, W.A., Nonaqueous solid particle dye dispersions, U.S. Patent 5,709,983 (1998).
3. Pugh, S.L.; Guthrie, J.T., Some characteristics of pigments that affect the kinetics of fading of prints made from water-based liquid ink formulations, *Dyes Pigments* **2002**, *55*, 109-121.
4. Lelu, S.; Novat, C.; Graillat, C.; Guyot, A.; Bourgeat-Lami, E., Encapsulation of an organic phthalocyanine blue pigment into polystyrene latex particles using a miniemulsion polymerization, *Polym. Int.* **2003**, *52*, 542-547.
5. Bishop, J.F.; Czekai, D.A., Ink jet inks containing nanoparticles of organic pigments, U.S. Patent 5,679,138 (1997).
6. Calvert, P., Inkjet printing for materials and devices, *Chem. Mat.* **2001**, *13*, 3299-3305.
7. Kinney, T.A.; von Gottberg, F.K.; Yu, Y.; Welch, J.F., Inkjet inks, inks, and other compositions containing colored pigments, U.S. Patent 6,506,245 (2003).
8. Chattopadhyay, P.; Gupta, R. B., Production of antibiotic nanoparticles using supercritical CO_2 as antisolvent with enhanced mass transfer, *Ind. Eng. Chem. Res.* **2001**, *40*, 3530-3539.
9. Mumper, R.J.; Cui, Z.; Oyewumi, M.O., Nanotemplate engineering of cell specific nanoparticles, *J. Disp. Sci. Technol.* **2003**, *24*, 569-588.
10. McKay, R.B., in *Technological Applications of Dispersions*, McKay, R.B.; Ed., Marcel Dekker, New York (1994) pp. 143-176.
11. Patton, T.C., *Paint Flow and Pigment Dispersion*, 2nd Edition, Wiley, New York (1979) pp. 376-512.
12. Scaringe, R.P.; Evans, S.; Van Hanehem, R.C., Co-milled pigments in ink jet ink, U.S. Patent 6,132,501 (2000).
13. Gutoff, E.B.; Swank, T.F. *AIChE Symp. Ser.* (No. 193) **1980**, *76*, 43-51.
14. Schwarz, C.; Mehnert, W.; Lucks, J.S.; Muller, R.H., Solid lipid nanoparticles (SLN) for controlled drug delivery. 1. Production characterization and sterilization, *J. Contr. Rel.*, **1994**, *30*, 83-96.
15. Auweter, H.; André, V.; Horn, D.; Lüddecke, E., The function of gelatin in controlled precipitation processes of nanosize particles, *J. Disp. Sci. Technol.* **1998**, *19*, 163-184.

16. Brick, M.C.; Palmer, H.J.; Whitesides, T.J., Formation of colloidal dispersions of organic materials in aqueous media by solvent shifting, *Langmuir*, **2003**, in press.

17. Texter, J., Organic particle precipitation, in *Reactions and Synthesis in Surfactant Systems*, Texter, J., Ed., Marcel Dekker, New York, (2001) pp. 577-607.

18. Diehl, D.R.; Factor, R.E., Solid particle dispersion filter dyes for photographic compositions, U.S. Patent 4,940,654 (1990).

19. Van Gilder and Langhorst, *ACS Symp. Ser.* No. 332, Am. Chem. Soc., Washington, DC (1987). Miller and Lines, *CRC Critical Reviews in Analytical Chemistry*, **20**, 2 (1988).

20. Stegeman, G.; van Asten, A.C.; Kraak, J.C.; Poppe, H.; Tijssen, R., Comparison of resolving power and separation time in thermal field-flow fractionation, hydrodynamic chromatography, and size-exclusion chromatography, *Anal. Chem.* **1994**, *66*, 1147-1160.

21. Texter, J., *A Practical Survey Course on Particle Characterization*, Strider Research Corporation, Rochester, New York, (2000) p. PS-7.

22. Silebi, C.A.; Dos Ramos, J.G., Method and apparatus for capillary hydrodynamic fractionation, U.S. *Patent* 5,089,126 (1992).

23. Dos Ramos, J. G., *Separation of Submicron Particles by Flow Fractionation: Capillary hydrodynamic fractionation (CHDF)*, University Microfilms, Ann Arbor, MI (1988), Order No. 8901867; Thesis, Lehigh University, Bethlehem, PA.

24. Venkatesan, J., *Particle Size Sensor Design and Application*, University Microfilms, Ann Arbor, MI (1992), Order No. 9312335; Thesis, Lehigh University, Bethlehem, PA.

25. Dos Ramos, J.G.; Silebi, C.A., Minimization of axial dispersion in hydrodynamic chromatography (HDC), *Polym. Mater. Sci. Eng*, **1986**, 54, 268-271.

26. Silebi, C.A.; Dos Ramos, J.G., Axial dispersion of submicron particles in capillary hydrodynamic fractionation, *AIChE J.* **1989**, *35*, 1351-1364.

27. Dos Ramos, J.G.; Silebi, C.A., An analysis of the separation of submicron particles by capillary hydrodynamic fractionation (CHDF), *J. Colloid Interface Sci.*, **1989**, *133*, 302-320.

28. Silebi, C.A.; Dos Ramos, J.G., Theoretical and experimental study of capillary hydrodynamic fractionation (CHDF, *Polym. Mater. Sci. Eng.* **1989**, *61*, 850-854.

29. Dos Ramos, J.G.; Silebi, C.A., Size analysis of submicrometer particles by capillary hydrodynamic fractionation (CHDF)", *Polym. Mater. Sci.* Eng. **1989**, *61*, 855-859.

30. Dos Ramos, J.G.; Silebi, C.A., Comparison between capillary hydrodynamic fractionation and more standard methods of particle size

analysis: Peak separation and broadening, *Polym. Mater. Sci. Eng.* **1989**, *61*, 860-864.

31. Dos Ramos, J.G.; Silebi, C.A., The determination of particle size distributions of submicrometer particles by capillary hydrodynamic fractionation (CHDF), *J. Colloid Interface Sci.* **1990**, 135, 155-177.

32. Dos Ramos, J.G.; Silebi, C.A., Submicron particle size and polymerization excess surfactant analysis by capillary hydrodynamic fractionation (CHDF), *Polym. Int.,* **1993**, *30*, 445-450.

33. Hollingsworth, A.D.; Silebi, C.A., Electrokinetic lift effects observed in the transport of submicrometer particles through microcapillary tubes, *Langmuir* **1996**, *12*, 613-623.

34. Belongia, B.M.; Baygents, J.C., Measurements on the diffusion coefficient of colloidal particles by Taylor-Aris dispersion, *J. Colloid Interface Sci.* **1997**, *195*, 19-31.

35. Elizalde, O.; Leal, G.P.; Leiza, J.R., Particle size distribution measurements of polymeric dispersions: A comparative study, *Part. Part. Sys. Char.* **2000**, *17*, 236-243.

36. Allen, T., *Particle Size Measurement,* Fourth Edition, Chapman and Hall, London (1990); pp. 399-401.

37. Oppenheimer, L.E., Interpretation of disk centrifuge data, *J. Colloid Interface Sci.,* **1983**, *92,* 350.

38. Coll, H.; Searles, C.G., Particle size analysis with the Joyce-Loebl disk centrifuge: A comparison of the line-start with the homogeneous-start method, *J. Colloid Interface Sci.* **1987**, *115,* 121.

39. Allen, T., *Particle Size Measurement,* Fourth Edition, Chapman and Hall, London (1990); pp. 413--414.

40. Texter, J., Capillary hydrodynamic fractionation of organic nanopigment dispersions, J. *Polym. Mater. Sci. Eng.* **2002**, *87,* 339-340.

41. Chmela, E.; Tijssen, R.; Blom, M.T.; Gardeniers, H.J.G.E.; van den Berg, A.; A chip system for size separation of macromolecules and particles by hydrodynamic chromatography, *Anal. Chem.* **2002**, *74*, 3470-3475.

Chapter 11

Determination of Complex Particle Size Distributions Using Packed Column Hydrodynamic Chromatography

E. Meehan and K. Tribe

Polymer Laboratories Ltd., Essex Road, Church Stretton, Shropshire SY6 6AX, United Kingdom

Packed column hydrodynamic chromatography (HDC) is a well established separation technique which fractionates particles according to their size. The practical range of particle size which can be separated by this technique is 5-3000nm. Using appropriate calibration procedures it has been shown that packed column HDC can be used to measure particle size distributions for a wide range of particle types. Commercial instrumentation for the application of this technology has recently been introduced. This article describes the design and operation of such instrumentation and illustrates the capabilities of the technique using a variety of particles which differ in composition, density and particle size.

Particle size can have a fundamental effect on the physical properties of colloidal dispersions. For many systems the measurement of average particle size is not sufficient, the presence of different size populations will have a strong influence on properties and could be related to the production process. Hydrodynamic chromatography (HDC), a technique for separating small particles by flow through a packed bed of non-porous particles, was invented by Small (1) and first described in 1974 (2). HDC provides a method for the separation of particles in suspension based on their size, eluting in the order largest to smallest. Based on this method of fractionation, HDC can then be applied to the determination of particle size for colloidal dispersions. Fractionation techniques, including packed column HDC, can offer advantages over non-fractionation techniques for particle sizing in that the method produces information about the average particle size and the distribution of particle size. Non-fractionation techniques are less well suited for the analysis of multi-modal samples or samples with broad particle size distributions due to the low resolution of the method. This paper describes the application of a new instrument, the PL-PSDA (Polymer Laboratories, UK), which can be used to determine the particle size distribution of a variety of colloidal dispersions based on the principle of packed column HDC.

Experimental

The PL-PSDA is an integrated, automated system in which the major components consist of a solvent delivery system, automated sample handling, separating cartridge and concentration detector. Instrument control, data acquisition, data analysis and reporting are all accessed from one PC based graphical user interface operating in a Windows environment.

In the PL-PSDA, a proprietary eluent (water based and containing a mixture of salts and surfactants at a controlled pH) is continuously pumped through the system at a constant flow rate of around 2.0 ml/min. A carousel based autosampler enables multiple sample vials to be loaded and sampled for continuous, unattended operation. Samples are prepared in 2 ml vials as a dilute slurry in the eluent after pre-filtration with 0.45µm or 2.0µm membrane filters, depending on the cartridge type employed. The sample under investigation, and a small molecule marker solution (3-nitrobenzenesulfonic acid), are introduced into the system via a two position, electrically actuated valve such that the eluent flow is not interrupted. At the heart of the system is a separating "cartridge", which consists of columns packed with non-porous polymeric particles of controlled particle size distribution, and the choice of cartridge dictates the particle size measuring range of the instrument with options available to cover the range from 5nm to 3um.

In the HDC technique described here, a suspension of particles flows through the cartridge which contains a packed bed of non-porous beads. The inter-particle spaces between the beads of the packed bed can be visualized as a series of small capillaries, within each capillary a velocity gradient exists described by a parabolic flow profile. The larger particles spend a greater proportion of time near the center of the capillary, experience the greater flow and are eluted from the column first The fractionation is therefore based on the different eluent velocity experienced by particles of different size due to the velocity gradient in the capillaries and solute particles elute from the system in order of decreasing size. Following the separation, sample components eluting from the cartridge are detected by a UV detector (λ=254nm) and in this technique detection occurs predominantly because the particles scatter the incident UV light, although there will be some contribution from absorption if the particles contain a UV chromaphore. The total analysis time is less than 10 minutes.

The detector response is used to calculate the concentration of particles of different size present in the sample and is corrected for Mie scattering using a detector response calibration curve. The subsequent computation of particle size also requires a calibration procedure which is specific to the separating cartridge employed. The primary system calibration procedure correlates particle size with retention factor, Rf, where Rf = Retention time marker / Retention time particle. The calibration plot is normally generated using 6 to 9 individual particle size standards and the data is fitted using a quadratic equation. Using a suite of specifically designed software algorithms, which include peak shape fitting and band broadening corrections, the instrument thus provides a complete particle size distribution for a sample as well as a calculated mean diameter and coefficient of variation (CV).

Universal Calibration

Cartridge calibration is usually performed using a series of certified polystyrene latex standards (Duke Scientific, Palo Alto CA, USA) to correlate Rf and particle size. Other types of certified particles have been analysed to illustrate the universality of the technique. Silica and melamine particles (G Kisker, Steinfurt, Germany), albeit in a more limited range of sizes, were analysed and the results for all three types of particle plotted on a "universal" calibration. The resultant calibration data in figure 1 shows that all types of particles lie on the same calibration line indicating that the separation method is independent of particle density (polystyrene = 1.05g/ml, melamine = 1.51g/ml, silica = 2.00g/ml). This observation suggests that the HDC technique described here has wide applicability for the determination of particle size distribution of many different particle chemistries.

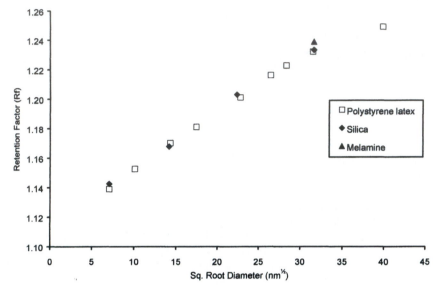

Figure 1. Particle size calibration data for different particle types with different particle densities.

Resolution in packed column HDC

The ability to separate particles of different size is predominantly controlled by the size of the beads in the packed bed of the separating cartridge. According to classical liquid chromatography theory, by decreasing the packed bed bead size, the chromatographic efficiency of the cartridge is increased (3). Also, with smaller packed bed bead size, the effective diameter of the inter-particle "capillaries" is reduced. Assuming a simple exclusion model, the ratio of solute particle radius to effective capillary radius, λ, determines to what extent the average particle velocity differs from the average solvent velocity (4). Therefore the packed bed bead size effectively controls the selectivity of the separation, or the ability to separate solute particles of similar sizes. The effect of packed bed bead size on HDC resolution is illustrated in figure 2 where the raw data chromatograms obtained for a series of five polystyrene latex particle size standards are overlaid and compared for two cartridges, Type 1 (smaller bead size) and Type 2 (larger bead size).

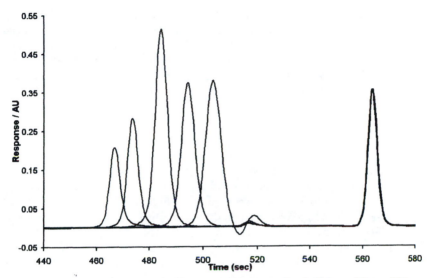

Figure 2a. Chromatograms of polystyrene latex standards (21nm, 50nm, 102nm, 204nm, 304nm) obtained using a smaller bed bead size Type 1 cartridge

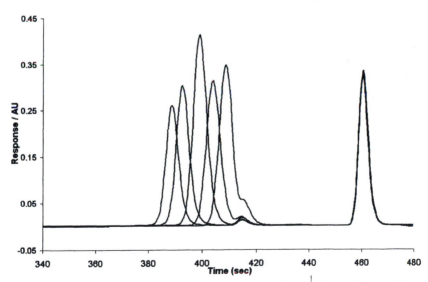

Figure 2b. Chromatograms of polystyrene latex standards (21nm, 50nm, 102nm, 204nm, 304nm) obtained using a larger bed bead size Type 2 cartridge

One limitation however is that as the inter-particle capillary diameter is reduced, the maximum size of particle that can pass through it is also reduced. Hence a small bead size cartridge Type 1 offers much greater resolution but over a more limited operating range (5-300nm) compared to a larger bead Type 2 cartridge (20-1500nm). As the upper size limit is approached, larger particles may not be completely recovered from the packed column and there is evidence of particle deposition in the cartridge.

Applications

Packed column HDC has been used to characterize a variety of particle types covering a wide range of size and distribution. The basic requirement for sample preparation is that the sample must be fully dispersed in the PL-PSDA eluent. Sample concentration is typically in the range 0.1-10.0mg/ml depending on the particle type and particle size range of interest.

Polymer latex particles, covering a wide range of size distribution and chemistry, have been extensively analysed. In many cases commercial samples of polymer latex contain a mixture of particles of different sizes and the high resolution of the HDC method is ideal for studying multi-modal distributions. This is illustrated in figure 3 which shows the particle size distribution determined for a mixture of equal quantities of three narrow dispersity polystyrene latex samples with nominal sizes of 102nm, 304nm and 519nm measured using the PL-PSDA fitted with a Type 2 cartridge.

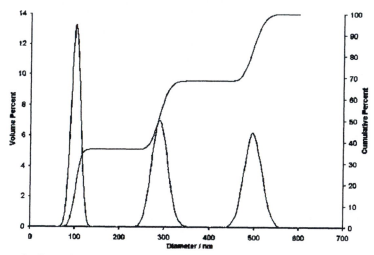

Figure 3. Particle size distribution obtained for an equal mixture of three polystyrene latexes (102nm, 304nm, 519nm).

The composition of the mixture of these three latex standards was varied in order to evaluate the ability of the technique to quantify the relative amounts of discrete populations in a multi-modal distribution. Table 1 compares the theoretical composition of four different mixtures with the results obtained by the HDC method.

Table I. Comparison of theoretical and calculated compositions of mixtures of three polystyrene latex standards (102nm:304nm:519nm)

Theoretical composition (%)	Calculated composition (%)
33 : 33 : 33	36 : 32 : 32
50 : 34 : 16	51 : 33 : 16
16 : 34 : 50	19 : 33 : 48
25 : 50 : 25	27 : 50 : 23

With this method it is possible to quantify with confidence the presence of a secondary species down to around 2-4% of the total volume of the sample. A further example of the effectiveness of this high resolution method is given in figure 4 which shows the particle size distribution measured for a commercial sample of PVC latex indicating the presence of three components varying in size, distribution and volume fraction in the sample.

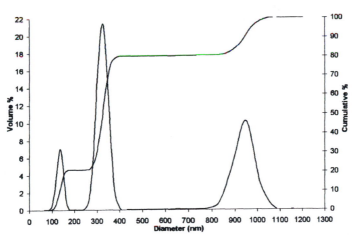

Figure 4. Particle size distribution determined for a commercial sample of PVC latex.

A further example of the application of packed column HDC is in the measurement of colloidal gold particles which are used extensively in clinical and diagnostic applications. The particle size is generally <100nm and the particle density is extremely high, at around 10g/ml. Figure 5 shows the particle size distribution of a sample of colloid gold in the presence of residual protein from an assay measured using a Type 1 cartridge for maximum resolution in the small particle size range (5-300nm).

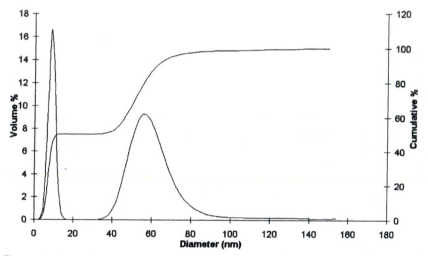

Figure 5. Particle size distribution for colloidal gold particles in the presence of residual protein.

Conclusions

The principle of packed column hydrodynamic chromatography has been successfully employed in a new instrument, the PL-PSDA, for the characterization of nano-particulate systems. The instrument measures the whole particle size distribution of the sample and thus exhibits information not necessarily observed by other techniques. The technique has wide applicability in terms of particle type and size distribution and the choice of cartridges permits optimization of resolution for specific applications.

References

1. H. Small, US Patent 3,865,717 (1975)

2. H. Small, *J. Colloid Interface Science*, **48**, 147 (1974)

3. In "Modern size exclusion liquid chromatography", WW Yau, JJ Kirkland, DD Bly, John Wiley Son, 1979, p86

4. G Stegeman, R Oostervink, J C Kraak, H Poppe, K K Unger, *Journal of Chromatography*, 506 (1990) 547-561

Chapter 12

Thermal Field-Flow Fractionation for Particle Analysis: Opportunities and Challenges

Paul M. Shiundu[1] and S. Kim Ratanathanawongs Williams[2]

[1]Department of Chemistry, University of Nairobi, P.O. Box 30197, Nairobi, Kenya (email address: pmshiundu@uonbi.ac.ke)
[2]Colorado School of Mines, Department of Chemistry and Geochemistry, Golden, CO 80401

The efficacy of thermal field-flow fractionation (ThFFF) for the retention and separation of submicrometer- and micrometer-size colloidal particles using both aqueous and nonaqueous carrier liquids is demonstrated. Several factors have been determined to influence retention in ThFFF. In these factors, lies the potential of ThFFF in particle analysis. For instance, the effects of chemical composition on retention are sufficiently strong that ThFFF can readily achieve the separation of equal-size particles of different composition, thereby opening up possibilities for compositional separations and analysis. Furthermore, the prospect of differentiating colloidal materials according to their surface composition is intriguing as it demonstrates the potential utility of ThFFF as a surface analysis tool. At first glance, the several factors that affect particle retention in ThFFF may seem complex and a nuisance to the general goal of developing ThFFF as a particle analysis tool. However, they offer significant opportunities and provide extra flexibility to the potential application of ThFFF in particle characterization. Results shown here demonstrate that varied levels of retention, separation resolution and speed among other separation features, can all be achieved through manipulation of various parameters such as field strength, cold wall temperature, ionic strength, and carrier composition. In this paper, we report and discuss these opportunities and their corresponding limitations. The potential of performing both bulk and surface compositional analysis of particles will be presented and the possibilities of ThFFF complementing both sedimentation and flow FFF techniques for the analysis of complex particulate materials will be highlighted.

Introduction

Thermal field-flow fractionation (ThFFF) is a member of the field-flow fractionation (FFF) family of chromatography-like separation techniques, which utilizes a temperature gradient as an external field. The thermal gradient induces differential retention of macromolecular and particulate species following differences in the physicochemical property of the sample components (1,2) by the phenomenon of thermal diffusion. ThFFF has traditionally been used for the separation and characterization of synthetic polymers (3,4) with the molecular mass range extending to ultra-high molecular mass polymers of mass ca. 20×10^6 (5). Recently however, the applications of ThFFF have expanded to include the retention and fractionation of particulate materials suspended in either aqueous or nonaqueous carrier liquids (6-10). A brief description of the retention theory of ThFFF is provided below.

Theory

The reader is referred to the literature (11, 12) for a more elaborate discussion of the mechanism and theory of retention in ThFFF. In brief, two transport processes govern the retention of a sample component in a ThFFF channel: The ordinary (mass) diffusion coefficient D and the thermal diffusion coefficient D_T. Particle properties are reflected both in D_T and D. The relationship between experimental retention time and particle properties is given by the following approximate equation (see reference (8) for derivation) for the ratio of retention time t_r to channel void time t^0

$$\frac{t_r}{t^0} = \frac{D_T \Delta T}{6D} \qquad (1)$$

where ΔT is the temperature drop applied across the channel. For spherical particles, D is related to particle diameter d by the Stokes-Einstein equation ($D=kT/3\pi\eta d$, where k is Boltzmann's constant, T is the absolute temperature in the region occupied by the sample component zone, and η is the viscosity of the carrier medium). Combining the Stokes-Einstein expression and equation 1 we obtain

$$\frac{t_r}{t^0} = \frac{\pi\eta d D_T \Delta T}{2kT} \qquad (2)$$

The ability of ThFFF to differentially retain and hence resolve particles of different sizes is related to the diameter-based selectivity S_d that can be expressed by

$$S_d = \frac{d \, log\left(t_r / t^0\right)}{d \, log \, d} \qquad (3)$$

It is conceivable that factors that affect t_r/t^0 would control S_d. However, lacking adequate theory to describe D_T, the matter of the dependence of S_d on various factors must be resolved by experimentation (see later).

The expression in equation 2 while based on various assumptions (6), illustrates the parameters that control retention in ThFFF. The thermal diffusion parameter D_T is quite complex and is the subject in part, of the work described in this article. Several factors have been empirically determined to affect D_T in various ways. Some of these effects have positive implications in the development of ThFFF as a complementary technique to both sedimentation and flow FFF techniques (1) for particle characterization. And this is the subject of discussion herein.

Experimental

The ThFFF system used for this work is similar in design to the Model T100 polymer fractionator from PostNova Analytics (salt Lake City, Utah) and the details are provided elsewhere (7). The 76 µm-thick channel spacer was confined between two chrome-plated copper bars. The top bar was heated using rods controlled by relay switches connected to a Microprocessor. The cold wall was cooled using continuously flowing tap water. Thermal sensors that were inserted into wells drilled into both the top and bottom bars monitored the temperatures of both the hot and cold walls. The channel breadth was 2.0 cm and the tip-to-tip length was 46cm. The carrier solutions (see figure captions for details) were delivered using a model M-6000A pump from Waters Associates (Milford, Massachusetts). A Model UV-106 detector from Linear Instruments (Reno, Nevada) operating at 240-nm wavelength was used to detect particles eluting from the channel. Twenty μl of sample was injected for each run using an injection valve (Model 7125, Rheodyne, Cotati, CA).

Samples and Carrier Solution

Standard polystyrene (PS) latex beads were obtained from Duke Scientific (Palo Alto, CA, USA) and Seradyn (Indianapolis, IN, USA). Polybutadiene (PB) and polydisperse micrometer-sized polymethylmethacrylate (PMMA) latex samples were obtained from Dow Chemical (Midland, NJ, USA) and Bangs Laboratories (Carmel, IN, USA), respectively. Core-shell latexes were contributions from AKZO (Arnheim, The Netherlands). The core is PS while the shells comprise of different ratios of butylacrylate (BuA) to methacrylic acid (MAA). These are designated as Latex 1 (BuA/MAA = 0:0, i.e., uncoated PS particle), Latex 2 (BuA/MAA = 7:4), Latex 3 (BuA/MAA = 9:2) and Latex 4 (BuA/MAA = 10:1). Both aqueous and nonaqueous carrier solutions were used and the details are given in figure captions (where applicable). In some instances (stated in figure captions where applicable), tetrabutyl ammonium perchlorate (TBAP) was used to modify the ionic strength of the carrier media.

Results and Discussion

The efficacy of ThFFF to separate latex particles according to size has been proven (6-9). Figures 1a and 1b illustrate the capability of ThFFF to separate aqueous suspensions of standard PS and PB latex particles, respectively, according to size. In this normal mode operation of ThFFF (1) the order of elution is from small to large particles. According to the theoretical expression given by equation 2, the retention time t_r in the normal mode FFF is expected to increase with particle diameter d, provided that D_T does not have a significant dependence on particle size. Both figures show the agreement between the experimental findings and theory. Figure 2 also shows that ThFFF can be used to separate larger particles in the micrometer-size range in an aqueous carrier medium. The particle size range covered in this figure is from 1.0 μm to 10.0 μm and comprises of eight sizes of PS latex spheres.

Since particle retention in the steric mode is not governed by Brownian motion but primarily by the effective diameter d (since d determines the particle's protrusion into the flow stream), the order of elution is reversed as shown in the

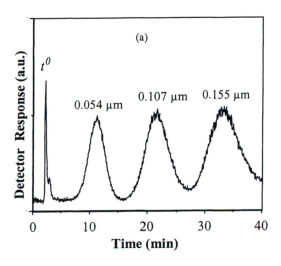

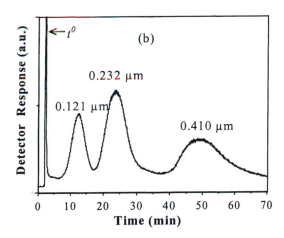

Figure 1. *Separation of submicron (a) PS latex particles at ΔT = 24 K and (b) PB latex particles at ΔT = 15 K suspended in an aqueous medium. Experimental conditions: [TBAP]= 1.0 mM, flow rate = 0.30 ml/min, and cold-wall temperature, T_c = 290 K.*

figure (13). Separation of particles suspended in a nonaqueous carrier solution in the steric mode has also been previously reported (7).

Several authors have identified various factors that influence retention of particulate matter in ThFFF (6-10, 14, 15), e.g., size, composition of both the particle and the suspending medium, ionic strength, cold wall temperature, and field strength. These parameters afford ThFFF extra flexibility in its potential role as a technique for the analysis of complex composite materials. For instance, subjecting a sample material to different ΔT conditions can effectively alter the level of retention of particles in ThFFF (and therefore the resolving power).

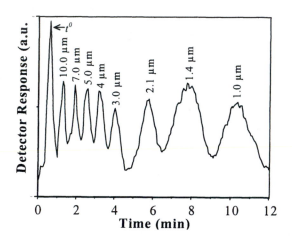

Figure 2. *Separation of micrometer-size PS latex particles in an aqueous carrier comprising of 0.1% (v/v) FL-70 surfactant and 0.02% NaN₃. Experimental conditions: [TBAP] = 1.0 mM, flow rate = 0.30 ml/min, ΔT = 75 K and cold-wall temperature, T_c = 305 K.*

According to equation 2, t_r of a species should increase in proportion to ΔT (provided that D_T is independent of temperature). An increase in ΔT will not only enhance particle retention and thus provide improved separation resolution, it will also prolong retention (unless offset by a higher flow rate) and thus can increase the experimental run-time beyond desirable levels. Figure 3, which is a log $(t_r/t^0))$ versus log d plot, shows the effect of ΔT on the retention of different size PS latex particles. The diameter-based selectivity, S_d values (a measure of the ability to differentially retain and therefore resolve particles according to their sizes) were determined to be 0.770, 0.830 and 0.880 corresponding to ΔT

values of 20K, 30K and 35K, respectively. These results signify a general increase in S_d with increases in ΔT and demonstrates the increased capability of ThFFF to differentially retain and thus resolve particles of different sizes. It should be noted that the criteria for preferred conditions could be either speed or adequate resolution between components (or both). However, undesirably long experimental run-times can be circumvented by higher flow rates (which would also have an influence on resolution). In addition, since the plots of log (t_r/t^0) versus log d are linear as shown in the figure, each individual plot can be used as a calibration line for the particle size distribution (PSD) of a polydisperse sample material of an identical composition. Hence the potential utility of ThFFF in PSD analysis.

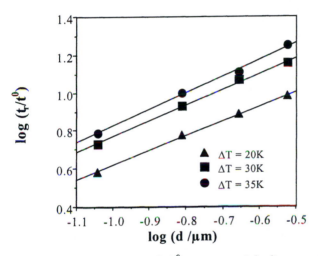

Figure 3. *Logarithmic plots of t_r/t^0 versus particle diameter d for PS particles suspended in acetonitrile (ACN). Experimental conditions: [TBAP] = 0.30 mM, flow rate = 0.29 ml/min, w = 76 μm and cold-wall temperature, T_c = 290 K.*

Further tunability in the operations of ThFFF with respect to particle retention is witnessed in the effect of cold wall temperature on retention. Figure 4 illustrates the effect of cold wall temperature in modulating the retention of PS latex particles in ThFFF. An aqueous carrier comprising of 0.1% (v/v) FL-70 surfactant and 1.0 mM concentration of tetrabutyl ammonium perchlorate (TBAP) was used with a corresponding ΔT of 24K. It is evident from this figure that lower cold wall temperatures yield longer retention times for both particle sizes, but with a larger effect for the plot corresponding to the larger size. These results however, contradict a previous study by Liu and Giddings (6) in which they reported increasing PS particle retention times with increasing cold wall temperatures in an aqueous medium. The only notable difference is that their aqueous carrier medium contained 0.02% sodium azide instead of 1.0 mM TBAP (in addition to the 0.1% FL-70 surfactant). The results of Figure 4 are further proof of the dramatic sensitivity of particle retention in ThFFF to the subtle changes in the additives of an aqueous carrier as has been previously reported by other workers (9, 15). Our results however, open up possibilities for studying temperature labile sample materials using ThFFF since relatively high ΔT conditions (hence increased levels of retention) can be achieved while maintaining favorable temperature conditions within the channel. The approach of lowering the cold wall temperature also ensures overall low power consumption.

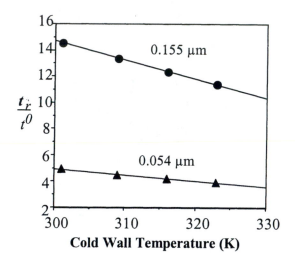

Figure 4. *Plots of retention time t_r (relative to void time t^0) versus cold wall temperature for 0.054 and 0.155 μm PS particles suspended in aqueous carrier solution containing 0.1% (v/v) FL-70 and 1.0 mM TBAP. Experimental conditions: Flow rate = 0.30 ml/min, w = 76 μm and ΔT = 24 K.*

Ionic strength has also been used to manipulate retention of particulate material in ThFFF (7, 9, 14, 15). Figure 5 shows the log (t_r/t^0) versus log d plots of PS latex particles suspended in aqueous carriers containing different concentrations of TBAP. Diameter-based selectivity, S_d values of 0.96, 1.40 and 1.50 were obtained with concentrations of TBAP corresponding to 0.05, 0.01 and 0.0050 mM, respectively; signifying the ability of ThFFF to differentially retain and resolve particles according to size with changes in ionic strength. However, it must be pointed out that an upper limit in ionic strength must be determined in order to circumvent problems associated with sample overloading and aggregation (9). Further results to show the effect of increased salt concentration on the retention and separation of similar-size 0.30 μm silica particles and 0.300 μm PS latex beads is shown in Figure 6. A ten-fold increase in the concentration of NaN$_3$ from 0.02% (w/v) to 0.20% (w/v) causes a significant change in the retention of the two particle populations. The figure also demonstrates the dependence of retention on chemical composition between dissimilar types of particles and hence the readily achieved separation of equal-size latex and silica particles.

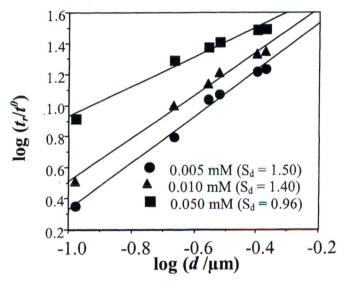

Figure 5. Logarithmic plots of t_r/t^0 versus particle diameter d for PS particles suspended in an aqueous carrier solution of varying concentration of TBAP. Experimental conditions: Flow rate = 0.30 ml/min, w = 76 μm and ΔT = 50 K.

This behavior also opens up possibilities for compositional separations and analysis by ThFFF. This retention dependence on chemical composition may however, also be interpreted as a drawback to the application of ThFFF in particle size distribution (PSD) analysis. For instance, in Figure 7, the fractogram of the standard PS latex particles should not be used to provide calibration constants for use in the determination of the PSD of the polydisperse PMMA latex sample material whose elution profile is also shown in the figure. Such a treatment would yield an incorrect PSD of the PMMA as 3.0 – 15.0 μm, contrary to the true PSD (1.0 – 10.0 μm, as provided by the manufacture and confirmed by scanning electron microscopy). This discrepancy arises because PS and PMMA are particles of different chemical compositions, a difference that would significantly cause differences in their retention behavior. Hence the need to use standards of identical composition as the sample material for calibration. This apparent drawback can however, be circumvented by coupling ThFFF system to an absolute size-detection system such as a multi-angle light scattering detector.

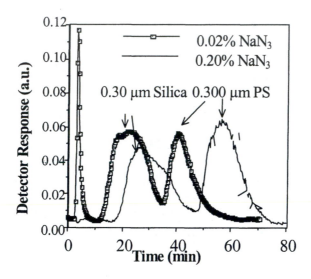

Figure 6. *Superimposed fractograms showing the separation of submicrometer populations of 0.30 μm silica and 0.300 μm PS particles suspended in an aqueous carrier medium containing 0.1% (v/v) FL-70 surfactant and two different concentrations of NaN₃. Experimental conditions: Flow rate = 0.30 ml/min, w = 76 μm and ΔT = 45 K.*

The true nature of the retention dependence on chemical composition is however, still a subject for further study. Our investigations indicate that both the bulk and surface properties of the particles contribute to the retention-composition behavior with the latter providing a more significant effect. Figure 8a shows superimposed elution profiles of core (latex 1) and core-shell latexes having different shell compositions. A phosphate buffer of ionic strength 0.01M and pH 9.0 (pH sufficient to induce swelling) was used as the carrier solution. Three significant observations can be made from these results. First, particle size is not the sole parameter influencing retention since the highly swelled core-shell latex 4 of overall diameter *ca.* 0.220 μm (10) is retained less than the PS core (latex 1 of diameter 0.135 μm).

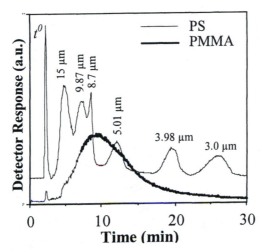

Figure 7. *Superimposed elution profiles of standard micrometer-size PS latex particles (light-solid line) and polydisperse PMMA sample (thick-solid line) suspended in aqueous carrier solution. Experimental conditions: [TBAP] = 1.0 mM, flow rate = 0.30 ml/min, ΔT = 24 K and cold-wall temperature, T_c =290 K.*

Second, the higher the BuA/MAA ratio of the shell (and hence the larger the degree of swelling), the shorter the retention time. In fact, the retention time can be used to obtain the shell composition of core-shell latex particles as shown in Figure 8b. Third, relatively larger changes in retention times with increasing BuA/MAA ratio are evident for higher ΔT conditions of 64 K than 37 K. This use of different field strengths also demonstrates the flexibility in the application of ThFFF to particle characterization.

The observation that the more swelled core-shell latexes are retained less than their non-swelled counterparts can only mean that the swelling has a counteracting influence on D_T since according to equation 2, t_r/t^0 increases in proportion to d. This implies that D_T decreases with increasing BuA/MAA ratio and hence the assertion that D_T depends on the chemical composition of the shell material of the core-shell latexes. It is however not clear whether this D_T - compositional dependence is governed by surface charge, hydration or some other factors.

The observed retention behavior is a manifestation of the influence of chemical composition on D_T. According to equation 2, t_r is proportional to particle diameter but it does not necessarily imply that larger particles will always be retained longer in the ThFFF channel irrespective of their chemical composition. This is further complicated by the fact that D_T is not independent of particle size (as is the case with polymers). Figure 9 shows the dependence of D_T on particle size as well as carrier composition. From these results, it is clear that the phenomenon of thermal diffusion is complex and poses great challenges while still offering possibilities for ThFFF in particle characterization.

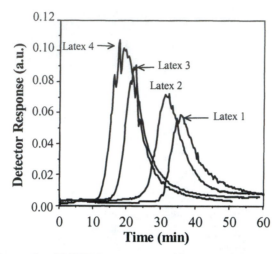

Figure 8a. *ThFFF fractograms of latex particles with different shell compositions (latex 1: BuA/MAA = 0:0, latex 2: BuA/MAA = 7:4, latex 3: BuA/MAA = 9:2 and latex 4: BuA/MAA = 10:1) obtained using 0.01 M phosphate buffer at pH 9.00. Experimental conditions: Flow rate = 0.20 ml/min, w = 76 μm, ΔT = 64 K and T_c = 301 K.*

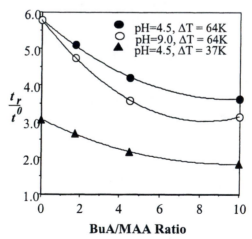

Figure 8b. *Calibration plots relating t_r/t^0 to the ratio of BuA/MAA obtained under different conditions of pH and ΔT. The cold-wall temperature, Tc = 301 K.*

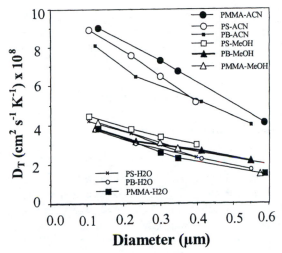

Figure 9. *Plots of thermal diffusion coefficient D_T versus particle diameter for PMMA, PB, and PS in acetonitrile (ACN), methanol (MeOH) and water (H_2O) carrier solutions. Experimental conditions: [TBAP] = 0.10 mM, flow rate = 0.30 ml/min, ΔT = 30 K and T_c = 287 K.*

Conclusion

This work demonstrates the ability of ThFFF to fractionate particulate matter according to size. ThFFF can be used to obtain size information of a sample material provided the availability of calibration standards of similar composition as the material under investigation or the coupling of the ThFFF system with an absolute particle size detector. Additionally, ThFFF is shown here to be sensitive to the shell composition of core-shell latexes and could therefore be utilized to obtain shell compositional information. Hence the potential of ThFFF in particle surface compositional analysis. The ready compatibility of the apparatus with both aqueous and nonaqueous carrier solutions affords the technique great flexibility. However, challenges still exist in determining what other factors affect D_T and the exact nature of such effects in order to make ThFFF an effective tool for separation and chemical characterization of particulate material.

References

1. Giddings, J.C. *Science* **1993**, 260, 1456.
2. Giddings, J.C.; Caldwell, K.D. In *Physical Methods of Chromatography;*

Rossiter, B.W.; Hamilton, J.F., Eds.; John Wiley: New York, NY, **1989**; Vol 3B, pp 867.

3. Thompson, G.H.; Myers, M.N.; Giddings, J.C. *Anal. Chem.* **1969,** 41, 1219.
4. Kirkland, J.J.; Rementer, S.W.; Yau, W.W. *J. Appl. Polym. Sci.* **1989,** 38, 1383.
5. Gao, Y.S.; Caldwell, K.D.; Myers, M.N.; Giddings, J.C. *Macromolecules* **1985,** 18, 1272.
6. Liu, G.; Giddings, J.C. *Chromatographia* **1992,** 34, 483.
7. Shiundu, P.M.; Liu, G.; Giddings, J.C. *Anal. Chem.* **1995,** 67, 2705.
8. Shiundu, P.M.; Giddings, J.C. *J. Chromatogr. A* **1995,** 715, 117.
9. Jeon, S.J.; Schimpf, M.E. In *Particle Size Distribution III. Assessment and Characterization*; Provder, T., Ed.; ACS Symposium Series; American Chemical Society: Washington, D.C., **1998**; Vol. 693, Ch. 13.
10. Ratanathanawongs, S.K.; Shiundu, P.M.; Giddings, J.C. *Colloids Surf. A: Physicochem. Eng. Aspects* **1995**, 105, 243.
11. Giddings, J.C. *Sep. Sci. Technol.* **1984,** 19, 831.
12. Giddings, J.C. *Chem. Eng. News* **1988,** 66, 34.
13. Williams, P.S.; Moon, M.H.; Giddings, J.C. In *Particle Size Analysis*; Staneleywood, N.G.; Lines, R.W., Eds.; Royal Society of Chemistry: Cambridge, UK, **1992**.
14. Jeon, S.J.; Schimpf, M.E.; Nyborg, A. *Anal. Chem.* **1997,** 69, 3442.
15. Mes, E.P.C.; Tijssen, R.; Kok, W.Th. *J. Chromatogr. A* **2001**, 907, 201.

Electrophoretic, Electrokinetic, and Acoustic Attenuation Methods

Chapter 13

New Insights into Latex Surface Chemical Heterogeneity from Electrophoretic Fingerprinting

R. L. Rowell, L. P. Yezek, and R. J. Bishop

Department of Chemistry, University of Massachusetts, Amherst, MA 01003

Electrophoresis is a sensitive probe of the surface electrical state of polymer latex particles. The recent approach of electrophoretic fingerprinting improves the electrical characterization and strongly reflects the chemical characterization. It is shown that the fingerprint patterns are useful in identifying surface chemistry, characterizing colloid stability and following time-dependent changes. New ideal chemical fingerprint patterns are proposed. The new approach is assessed by comparison with independent experimental work.

Introduction

Electrophoretic fingerprinting (EF), the idea of representing the measurable electrophoretic mobility of colloidal particles as a function of pH and $p\lambda$, the logarithm of the conductivity (microSiemens/cm is convenient) was conceived at the University of Massachusetts in 1986 (*1*). Since that time several papers on the fingerprinting approach have come from our laboratory (*2-11*) and collaborations (*12, 13*). A few other groups have taken up the approach. Marlow et al. have used EF to discern the biological activity of polystyrene latex microspheres that have been in contact with human body fluids (*14*). Paulke et al. have used EF to characterize latex particles with different surface groups

(*15*). Donath et al. have used EF to study the consecutive layer by layer polyelectrolyte adsorption onto charged polystyrene latex particles (*16*).

Our interest has been in exploring EF as a real-time tool to characterize any colloidal dispersion *in situ*, rapidly and non-invasively so as to be able to follow changes in the surface chemistry while reactions leading to those changes occur. In the present paper we have made some progress leading to the classification of strong acid, weak acid, zwitterionic and amphoteric oxidic surfaces. We know of no other direct probe such as what we seek.

Experimental Section

Materials. A monodisperse and ultra-clean, surfactant-free, carboxylic-acid-surface, polystyrene latex was obtained from Interfacial Dynamics Corporation, sample 10-317-82 as a 4.1% solids suspension in double-distilled water. Since the manufacturer's nominal diameter was 860 nm, we have labelled the sample C860. A similar latex sample from IDC had a carboxyl-amidine zwitterionic surface, batch 10-62-40, with a nominal diameter of 450 nm and was labeled CA450. A third dispersion in filtered double distilled water was titanium dioxide, the well-known P-25 from Degussa Corporation, predominantly anatase with an iso-electric point between pH 6 to 6.5. For short, we labelled the sample TDP25.

Electrophoresis. The dispersions were diluted to approximately 7.5×10^{-3} wt.% solids concentration (*17, 18*) using distilled water and various aqueous solutions of NaCl, NaNO$_3$, KCl, KNO$_3$, HCl, HNO$_3$, NaOH and KOH to obtain a matrix of concentrations giving a distribution of points in the pH, pλ plane. Samples were thoroughly mixed, equilibrated overnight and stabilized at 25°C before measurement. The average of several independent measurements of the electrophoretic mobility were obtained with a Brookhaven ZetaPlus and the solution pH and conductivity were measured at the same time.

Results

Figure 1 shows the electrophoretic fingerprints for C860 in four 1:1 salt systems using KCl, NaCl, KNO$_3$ and NaNO$_3$. Figure 2 shows the electrophoretic fingerprints for CA450 in the same salt systems. Figure 3 shows the electrophoretic fingerprints for TDP25 in the same salt systems. In all figures, the mobility is expressed in (μm/s)/(V/cm).

Discussion

The Present Results. The four patterns in Figure 1 are quite similar showing that the general pattern is characteristic of a particular colloid-salt system. The environment we have examined is inorganic 1:1 electrolytes. The results are reasonable since in the first approximation one would expect the 1:1 electrolytes to act as point charges. This is also borne out by similar trends shown in Figures 2 and 3 where, again, there is a similarity for 1:1 salts but distinctly different patterns for different colloids. This supports the idea of a characteristic fingerprint associated with a particular system.

If the surfaces in Figure 1 are indeed carboxyl as indicated by the supplier, then one characteristic of the carboxyl fingerprint is a dome maximum centered around pH 10, pλ3. A more precise location and detail of the maximum would require a survey at higher resolution than the scattering of points shown as small crosses for each sample measured. In a previous paper (*11*), we have shown that the dome maximum, the so-called anomalous electrokinetic maximum is largely a consequence of relaxation arising in the ionic diffuse layer surrounding the fully ionized carboxyls which give the macroion a large negative charge. The maximum mobility of -7 to -8 (μm/s)/(V/cm) corresponds to a zeta potential approaching -100mV.

A second characteristic of a carboxyl fingerprint is the line of zero mobility (LZM) that originates in a near-zero salt solution at a pH less than 3 and a pλ above 2. A simple carboxyl would not show an LZM. The mobility would go asymptotically to zero at the lowest attainable pH. This is the first observation of an LZM in a carboxyl latex. We suggest that the LZM may be attributable to protonation of the carboxyl group giving $-C(OH)^+_2$, a gem diol that would only be stable in solution. The LZM traces the pH-pλ dependence of the isoelectric point.

The general trend of the migration of the LZM is the same for each of the four salt systems. As pλ increases, the LZM moves toward higher pH. The effect is a regular drift toward higher pH for NaCl, KNO$_3$ and NaNO$_3$. But for KCl, there is a meander which could be an artifact.

The overall pattern is a practical definition of the conditions of colloid stability, which would be greatest in the pH-pλ domain of highest charge and least along the valley following the LZM.

Our patterns have been taken when the measurements are reproducible but we have observed pH drifts of as much as 1 unit overnight (*17*). Thus, with an apparatus built for the rapid collection of data using automatic titrators and high-speed measurement/storage, one could in principle follow the time-dependence of the fingerprints and thereby observe non-equilibrium interactions to various probe media and elucidate more understanding of the double layer from kinetic measurements.

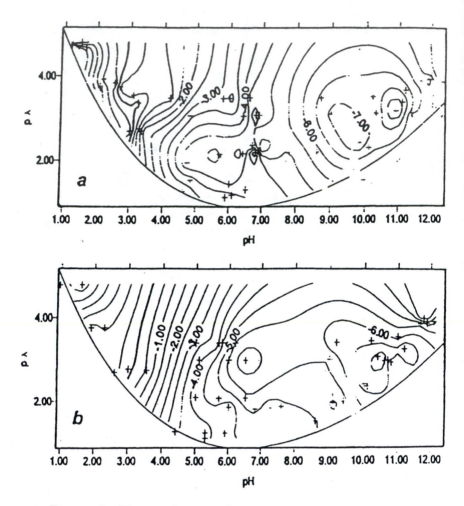

Figure 1. Electrophoretic fingerprints of carboxyl latex C860 in four electrolyte systems using (a) KCl, (b) NaCl, (c) KNO₃ and (d) NaNO₃ with the common-ion acid and base in each case.

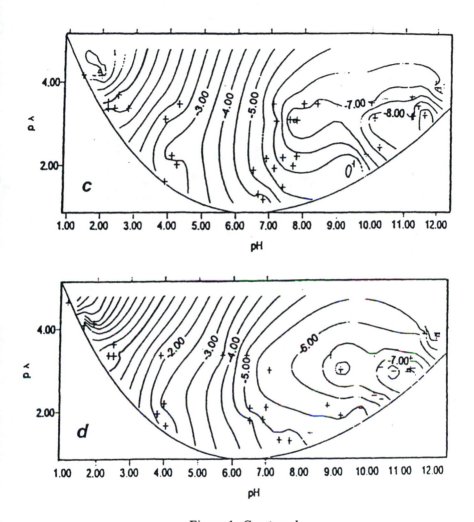

Figure 1. *Continued.*

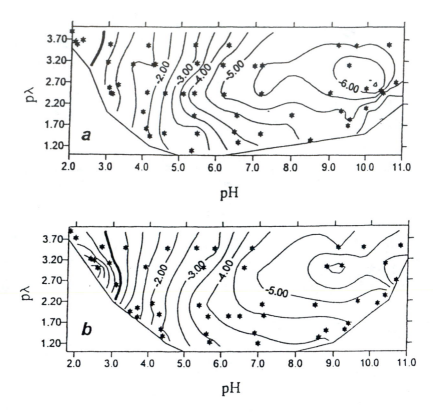

Figure 2. Electrophoretic fingerprints of carboxyl-amidine latex CA450 in four electrolyte systems using (a) KCl, (b) NaCl, (c) KNO₃ and (d) NaNO₃ with the common-ion acid and base in each case.

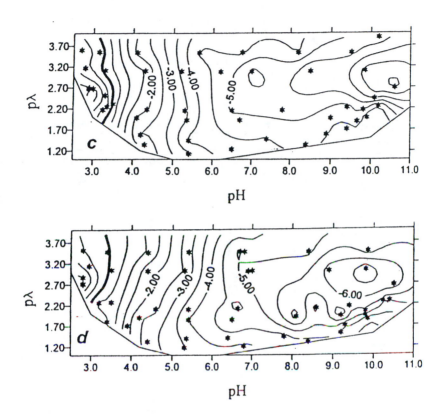

Figure 2. *Continued.*

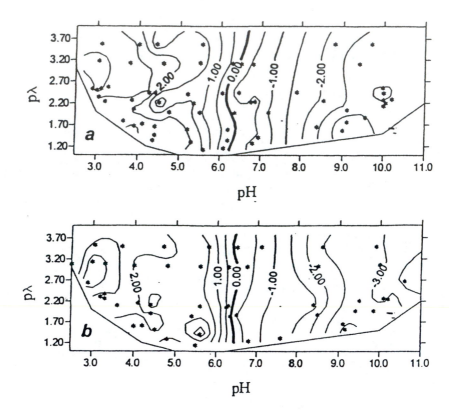

Figure 3. Electrophoretic fingerprints of titanium dioxide (Degussa P25) in four electrolyte systems using (a) KCl, (b) NaCl, (c) KNO₃ and (d) NaNO₃ with the common-ion acid and base in each case.

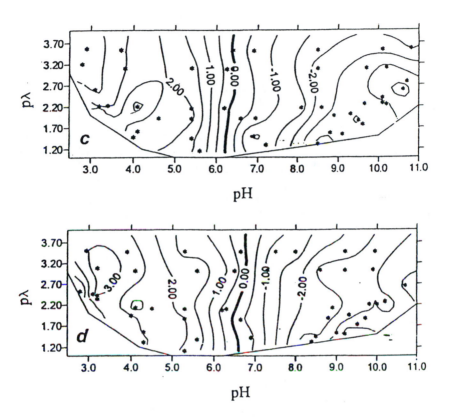

Figure 3. *Continued.*

One could extend the measurements to higher salt concentration which would be of interest in the study of biocolloids. For example, intravenous NaCl and Dextran-KCl-NaCl solutions are 150mM and 73mM respectively. We have examined only the low salt regime (below 30 mM) and have outlined the inaccessible region of pH-pλ space. A lower envelope blanking curve was obtained by polynominal fit to all our data and is given by (*17*):

$$p\lambda = 5.394 \times 10^{-4} \, pH - 0.02 \, pH^3 + 0.34pH^2 - 2.471pH + 7.115$$

The patterns for the carboxyl-amidine latex CA450 shown in Figure 2 show a different but generally similar pattern for the four salt systems. The pH-pλ coordinates of the carboxyl maximum are very nearly the same as Figure 1. A more critical evaluation of the differences would require a higher resolution of survey points. The pattern is reasonable well characterized with less than 50 samples. In an earlier study of a different carboxyl-amidine latex (*6*) we found little change in the general pattern for fingerprints generated using 38, 58, 73 and 106 samples. Our measurements lead us to believe that the surface was predominantly carboxyl (*18*).

The migration of the LZM from the CA450 systems in Figure 2 are different from that for the C860 systems shown in Figure 1. In NaNO₃, KNO₃ and NaCl the LZM shifts generally to lower pH at for CA450 vs. higher pH for C860. Furthermore, the LZMs originate at different pH-pλ coordinates on the envelope curves for the different latexes. This suggests that the LZM for the CA450 system is more likely controlled by the amidine group than arising from protonation of the carboxyl.

The patterns for the titanium dioxide dispersion TDP25 shown in Figure 3 are distinctly different from either Figure 1 or 2. The LZM shows little pH dependence and seems only slightly different for the different salt systems. There is no dome maximum in any regime and the absolute maximum mobility attained is less than half that found in the latex dispersions. Clearly, the oxide surface is distinguished by an amphoteric behavior of different symmetry and much lower mobility than the latex surfaces. In a previous paper (*11*) we have shown that the dome-maximum is a consequence of relaxation in the diffuse layer. The absence of a dome maximum in the TDP25 system is further confirmation. The ionization produced by the functional groups on the titanium dioxide surface does not provide the high surface charge and accompanying charge density in the diffuse layer required for relaxation to be a significant effect.

Comparison With Other Work. In an earlier paper on electrophoretic fingerprinting (*8*) of two surfactant-free sulfate latexes₁ of different diameters,

S686 and S1050, we found a dome maximum in the acid domain situated at pH 5, pλ 3, with a magnitude of −5 to −6. That work was independently carried out using a PenKem System 3000 (*19*). Comparison of the present work with the previous results shows that the sulfate and carboxyl latexes have distinctly different and widely separated characteristic dome maxima

The conclusion that the sulfate dome maximum lies in the fingerprinting domain at pH 5, pλ 3 sheds light on our earlier work on carboxyl latexes. Marlow and Rowell (*7*) compared experiment and theory to explain the electrophoretic fingerprint of IDC surfactant-free latex 10-32-14, a carboxyl latex C984. In order to obtain concordance between experiment and theory, they had to assume an expandable surface layer, a surface site density that varied with pλ and a pK$_a$ of 3.5 for the surface carboxyl. The electrophoretic fingerprint for C984 showed a dome maximum at pH 5, pλ 2 suggesting that the "carboxyl" latex had sufficient sulfate groups on the surface to dominate the electrophoretic properties. This casts doubt on the homogeneity of that latex and the previous theoretical interpretation.

The possibility that commercially available latexes may not have chemically uniform surfaces led us to examine the previous literature. There appears to be little analytical work carried out on the surface chemistry of surfactant-free latex preparations since the original work of Ottewill's group (*20*). They found that all of the latex particles from eleven independent preparations contained carboxyl, sulfate and hydroxyl groups. From our work with the TDP25 sample reported above we expect surface hydroxyls to play a much smaller role in electrophoresis than either carboxyl or sulfate. Also, the carboxyl groups will be more evident at high pH where the groups are fully ionized. When both sulfate and carboxyl groups are present, the sulfate groups may dominate the electrophoretic properties as appears to be the case with C984 discussed above.

Proposed Ideal Electrophoretic Fingerprints. Our studies to date suggest the possibility of several types of ideal electrophoretic fingerprints such as:

1. **Carboxyl (Weak acid).** We expect a single dome maximum centered around pH 10, pλ 3. An LZM would originate on the envelope of the experimental domain around pH 3 and shift toward higher pH as the pH and conductivity is increased due to deprotonization of the positively charged gem diol form of the carboxyl.
2. **Sulfate (Strong acid).** We expect a single dome maximum centered around pH 5, pλ 3 with no LZM in the pH-pλ domain.
3. **Amidine-carboxyl (Zwitterionic).** We expect the characteristic carboxyl dome maximum centered around pH 10, pλ 3 to dominate in

the alkaline regime and may well superimpose or mask any contribution from the negative form of the amidine ionization.

In the acid regime we would expect an LZM arising primarily from the protonation of the amidine group so that the LZM would shift little with an increase or decrease in pλ.

4. **Oxide (Amphoteric).** We would expect an oxidic surface to show a pattern similar to that of TiO_2 in Figure 3, i.e. no dome maximum in either the acid or alkaline regime, a low mobility and a symmetry about an LZM close to neutral pH that could shift with pH if specific ion adsorption were present. However, such trends assume an indifferent electrolyte and an iso-electric point close to neutrality. We would not be surprised to find that the patterns were influenced by non-indifferent electrolytes and iso-electric points that were farther removed from neutrality.

Summary and conclusions. There are several conclusions concerning the fingerprinting approach that may be drawn from the present work and the supporting references:

1. The fingerprint patterns are characteristic of the surface chemistry displayed in a particular medium. A carboxyl surface in a 1:1 electrolyte is expected to exhibit a dome maximum at pH 10, pλ 3 with an LZM originating at a pH less than 3 and a pλ above 2. Similarly, a sulphate surface is expected to exhibit a dome maximum at pH 5, pλ 3 with no LZM.

2. The migration of the valley of the LZM defines the region in space where the mobility is low and hence, systems at higher concentration are expected to be colloidally unstable.

3. Maximum colloid stability is expected at the highest mobilities, e.g., the region around the dome maximum for acidic surfaces.

4. The electrokinetic maximum is largely a consequence of relaxation.

5. The point charge approximation is a good approximation for simple 1:1 electrolytes.

6. For a system with changes in the surface chemistry, the fingerprinting method may be used to monitor and characterize the change without the need of detailed calculations based on an assumed model of the interface.

Acknowledgement. The authors are pleased to acknowledge partial support of this work which came from the Department of Chemistry, Olin Corporation, Brookhaven Instruments Corporation and the ONR Equipment Grant Program.

References

1. Morfesis, A.A. Ph.D. Thesis, University of Massachusetts 1986.
2. Morfesis, A.A. and Rowell, R.L. *ACS/PSME Proceedings*, **1986**, *54*, 514.
3. Rowell, R.L. in *Scientific Methods for the Study of Polymer Colloids*, Candau, F. and Ottewill, R.H. Eds.; Kluwer, The Netherlands 1990.
4. Morfesis, A.A. and Rowell, R.L. *Langmuir*, **1990**, *6*, 1088.
5. Rowell, R.L., Shiau, S-J. and Marlow, B.J., *ACS/PSME Proceedings*, **1990**, *62*, 52.
6. Rowell, R.L., Shiau, S-J. and Marlow, B.J. Ch 21 in *Particle Size Assessment and Characterization*, ACS Symp. Ser. 472, Provder, T. Ed., American Chemical Society, Washington, D.C. 1991.
7. Marlow, B.J. and Rowell, R.L. *Langmuir* **1991**, *7*, 2970.
8. Prescott, J.H., Shiau, S-J. and Rowell, R.L. *Langmuir*, **1993**, *9*, 2071.
9. Prescott, J.H., Rowell, R.L. and Bassett, D.R., *Langmuir*, **1997**, *13*, 1978.
10. Rowell, R.L., Bishop, R.J. and Yezek, L.P. Polyelectrolytes '98, Inuyama, Japan, 5/31-6/3/1998.
11. Yezek, L. and Rowell, R.L. *Langmuir*, **2000**, *16*, 5365.
12. Yezek, L., Rowell, R.L., Larwa, M. and Chibowski, E. *Colloids & Surfaces*, **1998**, *A 141*, 67.
13. Yezek, L., Rowell, R.L., Holysz, L. and Chibowski, E. *J. Colloid Interface. Sci.* **2000**, *225*, 227.
14. Marlow, B.J., Fairhurst, D., Schutt, W. *Langmuir*, **1988**, *4*, 776.
15. Paulke, B-R., Moeglich, P-M., Knippel, E., Budde, A., Nitsche, R. and Mueller, R.H. *Langmuir* **1995**, *11*, 70.
16. Donath, E., Walther, D., Shilov, V.N., Knippel, E., Budde, A., Lowack, K., Helm, C.A. and Moehwald, H. *Langmuir*, **1997**, *13*, 5294.

17. Bishop, R.J. Ph.D. Thesis, University of Massachusetts, 1999.
18. Yezek, L.P. Ph.D. Thesis, University of Massachusetts, 1999.
19. Shiau, S.-J. PhD. Thesis, University of Massachusetts, 1989.
20. Goodwin, J.W.; Hearn, J.; Ho, C.C.; Ottewill, R.H. *Br. Polymer J.* **1973**, *5*, 347.

Chapter 14

Pigment–Polyelectrolyte Interaction and Surface Modified Particle Characterization by Electrokinetic Sonic Amplitude Measurements

C. D. Eisenbach[1,2], Ch. Schaller[1], T. Schauer[1], and K. Dirnberger[2]

[1]Forschungsinstitut für Pigmente und Lacke e.V.,
D–70569 Stuttgart, Germany
[2]Institut für Angewandte Makromolekulare Chemie, Universität Stuttgart,
D–70569 Stuttgart, Germany

Detailed information about the interaction of copolymers with the pigment surface, e.g., adsorption/desorption phenomena as reflected from the dynamic mobility μ_D of the dispersed particle, can be obtained by applying the electrokinetic sonic amplitude (ESA) technique. With this method, a potential is measured from pressure waves generated by the movement of charged particles in an oscillating electric field. The variation of this potential with frequency can be used for determining μ_D of particles with a thin double layer and for calculating the zeta potential as well as the particle size. For fixed low frequencies as applied in this work, μ_D is effected by the surface charge of the particle only. The study of the change of the dynamic mobility μ_D of aqueous TiO_2 dispersion upon addition of amphipolar polyelectrolytes revealed that ESA is a powerful method to reveal the course of polymer-pigment interactions and how this is related to the molecular architecture of the employed polyelectrolytes. A model of initially competitive polymer micelle formation and mono-/bilayer polymer coating of the pigment surface is proposed.

For most technical applications of pigments and fillers a highly dispersed state of the particles is necessary in order to achieve optimal physical and application properties in the final coatings product. Stabilizers are employed in order to avoid agglomeration and flocculation of particles [1], and the principles of electrostatic, steric and electrostatic stabilization of pigments provide the theoretical basis for the design and optimal application of stabilizers [2,3]. Although in recent years polymer stabilizers have been proven to be efficient additives for high quality pigment concentrates and coating systems, the general understanding of interactions of the copolymers with the pigment surface is still limited. These interactions are influenced primarily by the surface properties of the pigment as employed and by the molecular architecture of the polymer molecule, and of course also by the nature of the dispersant medium.It is obvious that particularly in aqueous media the surface charge characteristics of the pigment as well as the amphipolar properties of the copolymer stabilizer mutually effect the particle/polymer/polymer interaction and thus the mechanism and efficiency of the particle stabilization. Consequently, analytical methods which are sensitive to changes in the surface charge density of particles are a promising tool for the study of copolymer-particle interaction and also for the comparative evaluation of the stabilization efficiency of different types of copolymers.

Here we report on our results obtained with the Electrokinetic Sonic Amplitude (ESA) method which is very sensitive to changes in the particle size and particle surface charge properties (cf. [4]). The ESA technique is based on an electroacoustic process and has become an increasingly important method to determine the surface characteristics like the dynamic mobility μ_D (electrophoretic mobility in an oscillating electric field) and the zeta potential of colloidal systems; ESA possess many advantages over more conventional electrokinetic techniques such as microelectrophoresis [5] and streaming potential [6, 7]. One major advantage of ESA is the capability to measure directly in concentrated colloidal systems (up to 40 % v/v) with particles ranging in size from a few nanometers to several micrometers.

Experimental

The synthesis and characterization of the amphipolar copolymers with regard to their structural composition and molecular weight characteristics has been described elsewhere [8].

Electrokinetic Sonic Amplitude (ESA) measurements were carried out with an Acoustosizer^{TM} AZR1, Colloidal Dynamics, Sidney, Australia, for 1 % volume per volume TiO$_2$ dispersion. All ESA data described in this paper relate to the dynamic mobility measurement at a frequency of 0.5 MHz. For the sake of better comparability of the various ESA data the so-called reduced dynamic mobility has been used; this is the ratio of the dynamic mobility μ of the particulate system after the addition of the copolymer and the initial dynamic mobility μ_0 of the pure pigment dispersed in water.

All dispersions were prepared at 23 °C according to DIN EN-ISO 8780-2 in water at pH 9 and conductivity of 0.05 S/m with a Scandex dispersing machine. This conductivity gives an initial salt concentration of 3.5×10^{-3} mole/l; due to addition of polyelectrolyte the total salt concentration increases up to 6×10^{-3} mole/l. The adjustment of pH and conductivity was done with 0.1 N HNO_3, 0.1 N KOH and 0.1 N KNO_3.

The Electrokinetic Sonic Amplitude Technique

In the context of this paper electroacoustics refers to two sorts of processes: when a sound wave passes through a colloidal system, a macroscopic potential difference is created (called the ultrasonic vibration potential or UVP). A first theoretical treatment of this effect was given by Debye [9]. The observed reverse effect that is observed when an alternating electric field is applied to a colloidal system generating a sound wave (called electrokinetic sonic amplitude or ESA effect) was discovered in the early 1980s [10]. A first theoretical treatment of the ESA effect was provided by O'Brien [11]. According to his theories the dynamic mobility can be used to calculate the zeta potential as well as the particle size even in concentrated colloidal systems [12-15].

Furthermore the ESA measurement is suited to give information about the particle/polymer interaction, e. g. adsorption/desorption phenomena. In earlier work, the adsorption of non-ionic polymers on the particle surface was investigated. It was found, that the presence of non-ionic polymers like poly(ethylene oxide) [16] or poly(vinyl alcohol) [17] on the particle surface causes an outward shift of the shear plane [18] that decreases the magnitude of the dynamic mobility. According to the gel layer theory, the frequency depending decrease of the magnitude of the dynamic mobility can be correlated with the thickness as well as the elasticity of the adsorbed layer of non-ionic polymers [19]. The lack of a theoretical background for the application of ESA to test the adsorption of polyelectrolytes makes it difficult to analyze the ESA data for this purpose.

For particles coated with polyelectrolytes no relationship between the dynamic mobility and the zeta potential as well as the particle size has been established so far. Nevertheless the ESA technique has been used in the past to investigate effects of polyelectrolytes in particulate systems. For example Rasmusson et al.

investigated the behavior of DNA in aqueous solution [20] and Walldal et al. correlated the dynamic mobility with flocculation of particles and the ionic strength [21]. The change in the zeta potential of particles upon coating with poly(acrylic acid) and poly(methacrylic acid) has been described for different nitrides [22] and oxides [23, 24]. In these cases the zeta potentials are only valid for a relatively flat conformation, i. e. for a thin layer of the polyelectrolyte on the particle surface, but this condition is only fulfilled at an alkaline pH [25].

A good correlation between the zeta potential [26] or the mobility spectra [27] as obtained by ESA measurements and adsorption isotherms has been reported in literature for polyelectrolytes. This infers that the ESA technique can be applied in principle to study the polymer adsorption in situ and thus to get an additional insight into the desorption/adsorption phenomena of polyelectrolytes. However, the situation complicates when employing amphipolar copolymers with non-adsorbing non-ionic as well as ionic structure elements: on the one hand the particle surface charge is altered as additional ionic groups are introduced with the polyelectrolyte adsorption, and on the other hand unpolar (non-adsorbing) structure elements cause a shift in the shear plane.

The suitability of the ESA measurement to detect changes in the surface charge density of the particles upon addition of polyelectrolytes which interact with the particle surface follows from the considerations of the limiting conditions from a theoretical point of view. This is shortly addressed below.

For a dilute suspension (particle volume fraction $\phi < 0.02$), it has been shown [11] that the ESA signal is given by Eq. 1:

$$\mathsf{ESA}(\omega) = \mathsf{A}(\omega) \cdot \varphi \cdot \frac{\Delta\rho}{\rho} \cdot \langle\mu_D\rangle \tag{1}$$

where ω is the angular frequency of the applied electric field, $A(\omega)$ is an apparatur constant, ρ is the density of the liquid, $\Delta\rho$ the density difference between the particles and the liquid and μ_D is the particle average dynamic mobility [28]. In the case of spherical particles with a thin double layer, μ_D is related to the zeta potential ζ by the following formula Eq. (2) [11]:

$$\mu_D = \frac{2 \cdot \varepsilon \cdot \zeta}{3 \cdot \eta} \cdot G\!\left(\frac{\omega \cdot a^2}{\nu}\right) \cdot [1 + f(\lambda, \omega)] \tag{2}$$

Here a is the particle radius, η is the viscosity of the liquid, ε is the permittivity of the liquid, and ν ($=\eta/\rho$) is the kinematic viscosity. The function $1+f(\lambda,\omega)$ in

Eq. (2) is proportional to the tangential component of the electric field at the particle surface [14]. The (dimensionless) quantity λ is a surface conductance parameter. The $G(\alpha)$ factor ($\alpha=(\omega a^2/\nu)$) represents the effect of the inertia forces which reduces the magnitude of the dynamic mobility μ_D, and which increases the phase lag as the frequency ω increases. This factor is given by the formula Eq. (3) [11]:

$$G(\alpha) = \frac{1+(1+i)\sqrt{\alpha/2}}{1+(1+i)\sqrt{\alpha/2}+i(\alpha/9)\cdot(3+2(\Delta\rho/\rho))} \qquad (3)$$

It is because of the inertia effect at higher frequencies that the particle size can be determined from the dynamic mobility by the ESA technique. At low frequencies ($\omega < 1$ MHz) and small particle size ($a < 1$ μm), α ($=(\omega a^2/\nu)$) is much smaller than 1, and thus $G(\alpha)$ reduces to 1; this means that the inertia forces of the particle can be neglected. Furthermore, for most water based colloidal systems $f(\lambda,\omega)$ becomes ½, and Eq. (2) reduces to the well known Smoluchowski equation, Eq. 4 (cf. [29]):

$$\mu_D = \frac{\varepsilon\cdot\zeta}{\eta} \qquad (4)$$

From Eq. (4) it is obvious that the dynamic mobility μ_D at low frequencies is only affected by the surface charge density of particles (as expressed by the zetapotential ζ). This means that the dynamic mobility μ_D as obtained by ESA measurement at 0,5 Hz is suited for the investigation of the particle/polymer interaction. At this low frequency the particle size and mass as well as the small mass increase due to adsorbed polymers (as compared to the mass of the naked particle) does not influence the dynamic mobility μ_D which is controlled by particle surface charge density. Considering the low polymer concentration of below 0.01 wt-% employed in the studies of the pigment/polyelectrolyte interactions, changes in the viscosity due to the added polymer that could effect the measured dynamic mobility value can be neclected.

For normalization and better elucidation of the pigment/polymer interactions, it is often advantageous to introduce the reduced dynamic mobility μ/μ_0; here μ stands for the mobility of particles with adsorbed polymer layer, and μ_0 represents the mobility of the non-treated particles (cf. [30]).

Experimental Results and Discussion

The molecular structure as well as schematic drawings of the ionic hydrophobic amphipolar polyelectrolytes are shown in Fig. 1.

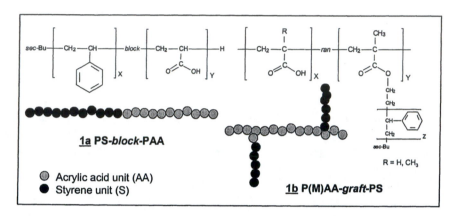

Figure 1 Molecular structure of the amphipolar copolyelectrolytes poly(styrene-block-acrylic acid) (PS-b-PAA; 1a) and poly((meth)acrylic acid-graft-styrene) (P(M)AA-g-PS; 1b) as well as the schematic cartoon models

The investigated ionic hydrophobic acrylic acid (AA)/styrene (S) based block and graft copolymers differed in the overall molecular weight, the block length and block length ratio, and in the length of the polymer backbone as well as the length of the grafts and the graft density. The controlled and well-defined structure of the block and graft copolymers is the basis for a comparative study of block and graft copolymers regarding the effect of changes in the constitution on the adsorption behavior.

For the investigation of the particle/polymer interaction, both kinetic (dynamic mobility μ_D vs time) as well as thermodynamic (dynamic mobility μ_D vs polymer concentration) ESA measurements are possible. The change of the reduced dynamic mobility μ_D/μ_0 of TiO_2 particles with time upon addition of different polyelectrolytes has already described elsewhere [30]. These mobility curves are indicative of a fast physisorption of the copolymers on the pigment surface due to hydrogen bonding, van der Waals forces and electrostatic interactions; similar interpretations have been given in the literature [31].

As reference and for the comparison of the data obtained with the amphipolar copolymers, first the adsorption behavior of the weakly negatively charged polyelectrolyte poly(acrylic acid) (PAA) on the TiO_2 surface was investigated by thermodynamic ESA measurements. Since the dynamic mobility μ_D at the low frequency of 0.5 MHz and for a small particle size ($\varnothing$ 370 nm for the used TiO_2) primarily depends on the change of the surface charge density, which is affected by the absorption of PAA, it is possible not only to obtain information about the coarse of PAA adsorption on TiO2 particles but also to determine the maximum adsorbed amount of PAA, designated as the so-called saturation concentration (SC). The change of the dynamic mobility μ_D with the concentration of added PAA is depicted in the Fig. 2a for different pH. Fig. 2b elucidates the determination of the SC which is given by the intersection of the extrapolated line of the limiting μ_D value and the extrapolated line of the initial slope of the change of μ_D upon polymer addition. For comparison, the adsorption isotherme as obtained by the sediment volume method is also included [30].

The obtained μ_D-polymer concentration dependencies in Fig. 2a are similar to adsorption isotherms of the langmuir type and correspond to the adsorption isotherms as obtained by the sediment volume method (s. Fig. 2b). It is obvious that the saturation concentration SC increases with decreasing pH. This pH depending adsorption behavior of polyelectrolytes is well known from the literature and is due to electrostatic repulsion between the charged particle surface and the fully dissociated PAA in solution [32]. The determind SC values are in good agreement with literature data: For example, the saturation concentration of 0.23 % w/w (2.3 mg/g) for PAA at pH 9 correlates well with results reported by Banash et al. [33]. The calculated surface area occupied by each carboxylic group of 0.88 nm^2 (based on a BET surface of 17 m^2/g and the molecular weight of 71,05 g/mole for an ionized acrylic monomer repeat unit of flatly adsorbed PAA) agrees well with the value of 0.86 nm^2 reported by Boisvert et al. which were obtained by conventional methods [34]. The determination of total organic carbon (TOC) of dried surface treated TiO_2 gave an adsorbed amount of 0.17 % w/w of PAA on TiO_2 (at pH 9) quite in accordance with the SC of 0.23 % w/w.

Although both the TiO_2 particle surface and the polyelectrolyte macromolecule become increasingly negatively charged with increasing pH, polyelectrolyte adsorption resulting in an overall increase in the surface charge density of the now polyelectrolyte-coated particle is still observed. This is because of the TiO_2 surface charge heterogeneity [35], i.e., the fraction of positive charges on the TiO_2 particle surface which, e.g., at pH 9 still amounts to about 0.3 positive charges per nm^2 (and 0.6 negative charges/nm^2) (cf. [34] and discussion in ref. [30]). These positively charged sites on the TiO_2 particle surface act as anchoring points for anionic carboxylate moieties of PAA. However, since only a fraction of the carboxylate groups of the polyelectrolyte polyanion is consumed in this ion pair formation, the overall charge density of the PAA-coated TiO$_2$ is higher than the charge density of the naked TiO$_2$ particle at the given pH; this is in accordance with the observed changes of the dynamic mobility.

As an intermediate result it can be concluded that the ESA technique can be regarded as a useful tool for in-situ investigation of polymer adsorption at a

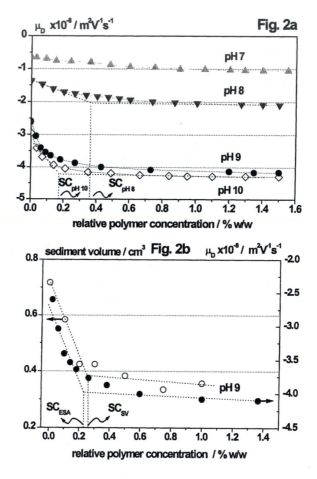

Figure 2 Dynamic mobility μ_D from ESA measurements for aqueous TiO$_2$ dispersions (1 % v/v) upon addition of poly(acrylic acid) (PAA) at different pH: 7 (▲); 8 (▼); 9 (●); 10 (◇) (s. Fig. 2a), and comparison of adsorption isotherm as obtained from sediment volume methode (○) with ESA method (Fig. 2b).

particle surface up to 40 % v/v concentration of particles, i.e., the dynamic mobility versus polymer concentration curves can be considered as "quasi" adsorption isotherms (cf. [27]).

After having discussed the interactions of PAA with TiO_2 particles, the adsorption behavior of poly(styrene-b-acrylic acid) (PS-*b*-PAA; <u>1a</u>) and poly((meth)acrylic acid-g-styrene) (P(M)AA-*g*-PS; <u>1b</u>) on the pigment surface is addressed. The adsorption behavior of amphipolar copolymer on the pigment surface is more complicated because association phenomena of the amphiphilic block copolymer including micelle formation have to be considered. Figure 3 shows the reduced dynamic mobility μ_D/μ_0 versus the relative copolymer concentration for PAA and the copolymer, i.e. for PS-*b*-PAA of different molecular constitution (Fig. 3a) and for PMAA-*g*-PS and PS-*b*-PAA samples of comparable molecular weight and similar overall comonomer composition (Fig. 3b).

The plots of the reduced dynamic mobility μ_D/μ_0 versus the relative copolymer concentration (Fig. 3a) show that the adsorption behavior of the amphipolar block copolymer is quite different as compared to the pure PAA: whereas for the case of the PAA adsorption an increase in the reduced dynamic mobility μ_D/μ_0 is observed from the beginning, corresponding to an increase in the overall surface charge density of the coated pigment, such a behavior is seen only for block copolymers of relatively low overall molecular weight and also short polystyrene block length. If the length of the hydrophobic block is increased, first a decrease in the reduced dynamic mobility is observed until a minimum is reached (region I); this decrease is correlated with the PS block length. After a minimum has been passed, the mobility increases towards a limiting value (region II) and reaches a saturation concentration (region III). For a given polystyrene block length, the increase is somewhat higher if the poly(acrylic acid) block is larger. This effect substantiates the influence of the ionic blocks on the overall particle charge as well as on the saturation concentration SC. The relationship between the polymer block and graft length and particle mobility have been described in more detailed elsewhere [30,36,37]. Phenomenologically, the same observations have been made with graft copolymers of comparable molecular weight range and overall comonomer composition (s. Fig. 3b). Similar results that the dynamic mobility μ_D is only affected by the charge density of the copolyelectrolytes and not by the molecular architecture have been reported in the literature as well [38].

Regarding the dependency of the reduced dynamic mobility μ_D/μ_0 on the concentration of an added copolymer, it is interesting to compare the critical micelle concentration (CMC) of the amphipolar copolymers in pure water and in

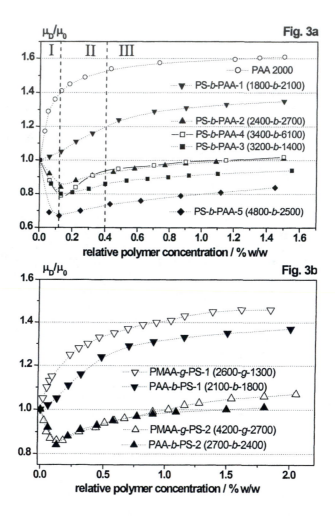

Figure 3 Reduced dynamic mobility μ_D/μ_0 from ESA measurements for aqueous TiO_2 dispersions (1 % v/v, pH 9) upon addition of poly(styrene-block-acrylic acid) (PS-b-AA; 1a) with different constitution (s. Fig. 3a) and of poly(methacrylic acid-graft-styrene) (PMAA-g-S; 1b) and poly(styrene-block-acrylic acid) (PS-b-AA; 1b) with comparable molecular weight range and overall comonomer composition

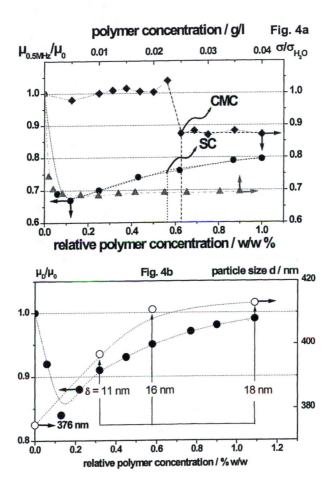

Figure 4 Change of the reduced surface tension σ/σ₀ of water (▲) and aqueous TiO₂ dispersion (◆) (0.25 % v/v, pH 9) upon addition of poly(styrene-b-acrylic acid) (PS-b-PAA-5) and comparison with the change in the reduced dynamic mobility μ_D/μ₀ (●) (TiO₂ dispersion 1 % v/v, pH 9) (Fig. 4a: CMC: critical micelle concentration, SC: saturation concentration); particle size d and polymer layer thickness δ of TiO₂ particles with adsorbed PS-b-PAA-2 (○) (0.25 % v/v, pH 9) compared to the reduced dynamic mobility μ_D/μ₀ (●) (1 % v/v, pH 9), Fig. 4b

aqueous TiO_2. Figure 4a shows the reduced surface tension σ/σ_0 of the aqueous poly(styrene-*block*-acrylic acid) (PS-*b*-PAA-5; 1a)/TiO_2 system, where σ_0 represents the surface tension of the pure aqueous TiO_2 dispersion and σ is the surface tension of the TiO_2 dispersion with added copolymer (right ordinate). The corresponding reduced dynamic mobilities μ_D/μ_0 (left ordinate) are given as well. Examples of how the particle size and the polymer layer thickness change due to block copolymer adsorption (PS-*b*-PAA-2; 1a) on the surface of TiO_2 particles are also given together with the corresponding reduced dynamic mobility μ_D/μ_0 data (Fig. 4b).

Fig. 4a shows that in the low concentration region below the saturation concentration (for SC compare Fig. 3) the added block copolymer adsorbs onto the particle surface; no change in the reduced surface tension σ/σ_0 occurs since all the added copolymer is consumed by the adsorption process. For comparison, the surface tension of the TiO_2-free water drops very sharply after little addition of PS-b-PAA 5: the CMC of PS-*b*-PAA-5 in aqueous solution is about 0.5×10^{-3} g/l which is located below the region I (cf. [28]). It is only after the saturation concentration (SC) has been exceeded that the reduced surface tension σ/σ_0 decreases. The CMC of the PS-b-PAA-5 is reached at a much higher copolymer concentration in the case of the TiO_2 dispersion as compared to the pure water medium: the concentration difference reflects the amount of copolymer adsorbed at the TiO_2 surface, and only when the TiO_2 surface is covered by copolymer, micelle formation of excess copolymer molecules can occur. Consequently, the CMC in the TiO_2 dispersion corresponds to the saturation concentration as determined by ESA.

From Fig. 4b it is obvious that the particle size d and the layer thickness δ of the adsorbed polymer increase with increasing block copolymer concentration and remain constant only after the saturation concentration has been slightly exceeded. This finding is in accordance with the CMC measurements which have shown that formation of free micelles only occurs after the deposition of amphipolar copolymer on the TiO_2 surface is completed.

These results, i.e. the good correlation between CMC values, the saturation concentration (SC) and thickness of the adsorbed polymer layer are complementary to each other and reflect the applicability of the ESA technique for the in-situ analysis of the pigment/polymer interactions, e. g., the polymer adsorption.

The proposed adsorption mechanism of amphiphilic copolyelectrolytes is shown in the schematic Figure 5.

The initial adsorption of the block as well as of the graft copolymer is driven by the interaction of the PAA block (graft) with the pigment surface. The copolymers are attached with the PAA part to the TiO_2 surface whereby the positively charged sites on the overall negatively charged TiO_2 particle act as anchoring points for carboxylate groups (region I). The PS part of the attached amphipolar copolymer pointing to the aqueous phase causes a shift of the shear plane resulting in a decrease of the zeta potential as reflected from the decrease in the dynamic mobility (see Fig. 3 and 4).

After the formation of a sort of a copolymer monolayer which is indicated by the minimum in the ESA measurement curves (see Fig. 3 and 4), a bilayer-like coating is formed by hydrophobic-hydrophobic intermolecular interaction of the hydrophobic PS sites on the TiO_2 particle with the PS part of further added copolymer (region II and III). This again alters the shear plane, but now in the opposite direction due to the ionic carboxyl groups of the PAA part now extending into the aqueous phase; this means an increasing zeta potential, i.e., the negative charge density of the TiO_2 particle is increased, and thus is the dynamic mobility.

Once this bilayer formation is completed which corresponds to the saturation concentration SC (see Fig. 4a), excess addition of copolymer leads to micelle formation which may be further adsorbed as so-called solloides (cf. [30]).

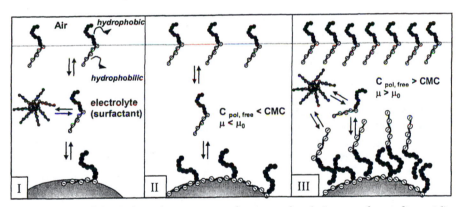

Figure 5 Model of the adsorption mechanism of poly(styrene-b-acrylic acid) (PS-b-PAA) on TiO_2 pigment in aqueous medium (I) Micelle/unimer equilibrium at concentration above cmc and onset of copolymer adsorption (II) depletion of the aqueous phase of copolymer and formation of monomolecular polymer layer on the pigment surface (III) bilayer formation in the higher concentration regime via intermolecular hydrophobic-hydrophobic interactions until the saturation concentration is reached, followed by micelle formation of excess block copolymer molecules

Conclusions

Starting from theoretical considerations that effects of inertia forces on the dynamic mobility μ of dispersed particles become neglectably for small particles

< 1 µm at measuring frequencies ω below 1 MHz, the ESA method has been considered to be suited to follow the polymer adsorption on the surface of small particles. This has been proven to hold true for polyelectrolytes such as poly(acrylic acid) as well as amphipolar acrylic acid/styrene based block and graft copolymers. The ESA data correlate well with conventional adsorption isotherms; complementary CMC and particle size measurements infer a sequential adsorption mechanism. Thus the ESA method can be regarded as a useful tool for the investigation of particle/polyelectrolyte interactions and polyelectrolyte adsorption for particle concentrations up to 40 % v/v.

Acknowledgement

Financial support of this study by the Deutsche Forschungsgemeinschaft (Forschungsschwerpunkt Polyelektrolyte, Grant No. Ei 147/19-2) is gratefully acknowledged.

References

1. Clayton, J.; Pigm. Resin Technol. **1998**, 27, p 231.
2. Tadros, Th.; Solid/Liquid Dispersions, Academic Press, London 1987.
3. Napper, D. H.; Polymeric Stabilization of Colloidal Dispersions, Academic Press, New York 1983.
4. Carrasso, M. L.; Rowlands, W.N.; O'Brien R.W.; J. Coll. Interf. Sci. **1997**, 193, p 200.
5. Hunter R. J.; Introduction to Modern Colloid Science, Oxford University Press, Oxford, New York, Melbourne 1983.
6. Jacobasch, H.-J.; Oberflächenchemie Faserbildender Polymere, Akademie-Verlag, Berlin **1984**.
7. Osterhold, M.; Schimmelpfennig, K.; Farbe&Lack **1992**, 98, p 841.
8. Schaller, Ch.; Schauer, T.; Dirnberger, K.; Eisenbach, C.D.; Eur. Phys. J. E **2001**, 6, p 365.
9. Debye, P. J.; J. Chem. Phys. **1933**, 1, p 13.
10. Oja, T.; Petersen, G. L.; Cannon, C. D.; U.S. Patent 4.497.208, 1985.
11. O'Brien, R. W.; J. Fluid Mech. 190 (1988) 71
12. Loewenberg, M.; O'Brien, R. W.; J. Col. Int. Sci. **1991**, 150, p 158.
13. Rider, P. F.; O'Brien, R. W.; J. Fluid Mech. **1993**, 257, p 607.
14. O'Brien, R. W.; Cannon, D. W.; Rowlands, W. N.; J. Col. Int. Sci. **1995**, 173, p 406.
15. O'Brien, R. W.; Rowlands, W. N.; Hunter, R. J.; Ceram. Trans. **1994**, 54, p 53.
16. Maier, H.; Baker, J. A.; Berg, J. C.; J. Col. Int. Sci. **1987**, 119, p 512.
17. Miller, N. P.; Berg, J. C.; Col. Surf. **1991**, 59, p 119.

18. Lyklema, J.; Fundamentals of Interface and Colloid Science II, Solid-Liquid Interfaces, Academic Press, London 1995.
19. Carasso, M. L.; Rowlands, W. N.; O'Brien, R. W.; J. Col. Int. Sci. **1997**, 193, p 200.
20. Rassmusson, M.; Åkermann, B.; Langmuir **1998**, 14, p 3512.
21. Walldal, C; Åkermann, B.; Langmuir **1999**, 15, p 5237.
22. Paik, U.; Hackley, V. A.; Lee, H.-W.; J. Am. Ceram. Soc. **1999**, 82, p 833.
23. Costa, L.; Galassi, C.; Greenwood, R.; J. Col. Int. Sci. **2000**, 228, p 73.
24. Beattie, J. K.; Djerdjev, A.; J. Am. Ceram. Soc. **2000**, 83, p 2360
25. Greenwood, R.; Bergstrom, L.; J. Europ. Ceram. Soc. **1997**, 17, p 537.
26. Pettersson, A.; Marino, G.; Pursiheimo, A.; Rosenholm, J. B.; J. Col. Int. Sci. **2000**, 228, p 73.
27. Walldal, J. Col. Int. Sci. **1999**, 217, p 49.
28. Hunter, R. J.; O'Brien, R. W.; Colloids Surfaces A: Physicochem. Eng. Asp. **1997**, 126, p 123.
29. Hunter, R. J.; Colloids Surfaces A: Physicochem. Eng. Asp. **1998**, 141, p 37.
30. Schaller, Ch.; Schauer, T.; Dirnberger, K.; Eisenbach, C.D.; Europ. Phys. J., **2001**, 6, p 365.
31. Lee, D. H.; Condrate, R. A.; Reed, J. S.; J. Mat. Sci. **1996**, 31, p 471.
32. Gebhardt, J. E.; Fürstenau, D. W.; Col. Surf. **1983**, 7, p 221.
33. Banash, M. A.; Croll, S. G.; Prog. Org. Coat. **1999**, 35, p 37.
34. Boisvert, J.-P.; Persello, J.; Castaing, J.-C.; Cabane, B.; Colloids and Surfaces A: Physicochem. Eng. Asp. **2001**, 178, p 187.
35. Hoogveveen, N.G.; Cohen-Stuart, M.A.; Fleer, G.J.; J. Col. Interf. Sci. **1996**, 133, p 182.
36. Schaller, Ch.; Schauer, T.; Dirnberger, K.; Eisenbach, C:D:; Farbe+Lack **2001**, 11, p 58.
37. Schaller, Ch.; Schauer, T.; Dirnberger, K.; Eisenbach, C.D.; Prog. Org. Coat., to be published
38. Walldal, C.; Åkermann, B.; Langmuir **1999**, 15, p 5237.

Chapter 15

Acoustic Analysis of Concentrated Colloidal Systems

Tõnis Oja[1], J. Gabriel DosRamos[1], and Robert W. Reed[2]

[1]Matec Applied Sciences, 56 Hudson Street, Northborough, MA 01532
[2]Physics Department, East Stroudsburg University,
East Stroudsburg, PA 18301

Introduction

In this paper, we give a brief review of a class of measurements that can be
made using high frequency (1-100 MHz) acoustics. The emphasis will be on
measurements carried out on concentrated liquid-borne particle systems. In
particular, we will be interested in colloidal systems, i.e. systems of fine (< 20
microns) particles or liquid droplets, suspended in liquids. Since acoustic waves
propagate in all material media, acoustic measurements have been made in a
vast variety of gaseous, liquid, and solid systems. The field of acoustic
measurements is immense. See, for example, Reference 1 dealing with
ultrasonics in fluid characterization.
The advent of high-speed computing has permitted rapid and precise
measurements of acoustic attenuation, sound speed, and acoustic impedance
over a large range of frequencies. With data quickly available, it is now possible
to compute, in concentrated dispersions, the mean particle size and size
distributions, the concentration and viscosity in real time and thus be able to
follow the evolution in time of an undiluted, hence more realistic colloid system.
Rapid sound-speed measurements compliment the attenuation results and help
speed the analysis. To date, most particle characterization of dispersed systems
has been carried out primarily by laser light scattering. Light scattering methods
require the severe dilution of the dispersed system and hence do not permit to
study aggregation, flocculation, gelling and other phenomena associated with
real systems.

Analysis Method

As described by Reed *et. al.* (2), acoustic attenuation measurements require the
very precise measurement of three parameters: ultrasonic echo amplitudes, the

fluid path length, and time. The APS - 100, instrument from Matec Applied Sciences (3) performs these measurements over 1-100 MHz frequency range. Samples are normally stirred or pumped during the measurement to ensure homogenous mixing, and to prevent particle settling. Particles in size range from 10 nm to 100 microns can be measured.

The fluid path length is measured by using a high-precision motion stage. The APS – 100 maximizes the accurate reflector position measurement by simultaneously measuring the sound speed in the sample at multiple reflector locations, and correcting any minor position errors (4).

The APS measures echo amplitude and sound speed by using a reflector based design allowing the comparison of the transmitted and reflected sound wave using the same transducer and electronics. This permits use of coherent tone bursts. As a result, the APS can perform many repetitive measurements for optimal signal averaging in order to maximize resolution, accuracy and reproducibility. The advantage of coherency is lost when one transducer emits the sound wave and a different transducer receives it.

Additionally, the APS reflector design eliminates the need for O-rings, through which a transducer moves.

Acoustic attenuation must be measured at multiple spacing for two reasons: (1) high frequency measurements have higher attenuation so they have to be made at short paths, whereas at low frequencies, longer path lengths are required due to much lower attenuations; and (2) the attenuation vs. frequency curve must be built with as many data points as possible in order to produce reliable particle size distribution data.

Results

Figure 1 shows results of the acoustic attenuation measurements of pure water for eight different path lengths. Attenuation is simply defined as **Attenuation Log_{10} (I_i/I_o)** (in decibels-dB). I_o and I_i are the measured incident and out - going sound intensities.

It is apparent that all attenuation vs. frequency curves are parabolic, that is, the attenuation is proportional to the frequency squared.

A more demanding test of the goodness of the measured data is obtained by plotting the above results on a log-log scale, shown in Figure 2. It is evident that the linear scales mask the uncertainties of the measured attenuation at low attenuation values. We have concluded that any measured attenuation less than one db is not sufficiently reliable and is discarded from any data set.

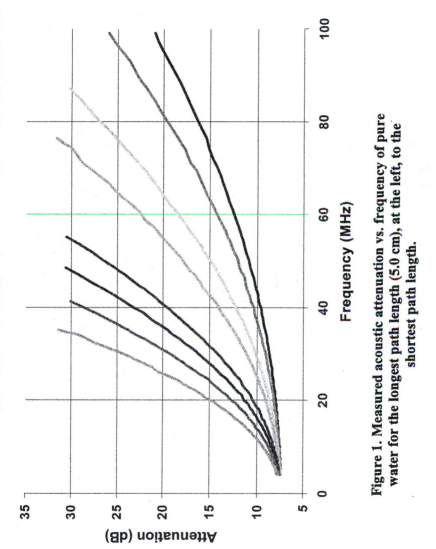

Figure 1. Measured acoustic attenuation vs. frequency of pure water for the longest path length (5.0 cm), at the left, to the shortest path length.

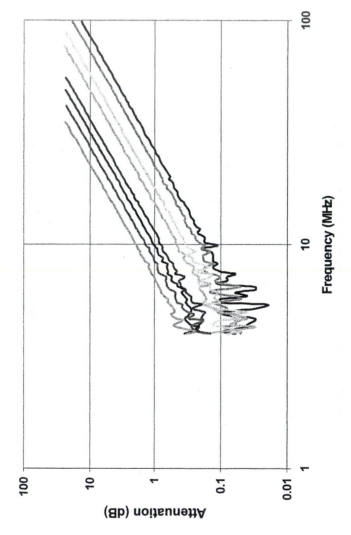

Figure 2. Measured attenuation vs. frequency of pure water plotted on log-log scales

Combining the attenuation results (in db) from the eight different path lengths into one data set by computing the attenuation per unit distance (db/cm), we obtain the graph shown in Figure 3. On the log-log scale, the data fall on a straight line of slope = 2. This is the beginning point of all acoustic attenuation measurements in water as well as in other liquids. In this frequency range, one must be able to generate a straight-line plot on log – log scales if there is to be any hope of obtaining good particle sizing results. In a Newtonian ("ideal") liquid, the slope = 2.00. In non-Newtonian liquids, the slope is less than 2.00.

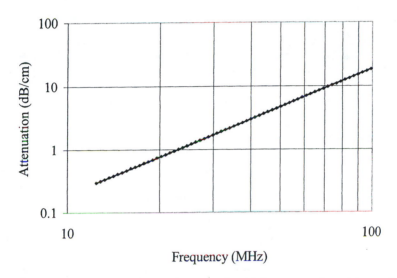

Figure 3. Attenuation of water in db/cm as a function of frequency.

In Figure 4 are shown the attenuation results measured in water as well as in 100 nanometer (NP 1040) and 300 nanometer (NP 3040) silica dispersions. It is important to note that as the particle size increases, the resultant attenuation curves deviate increasingly from the linear, pure water curve. It is the curve shape that is important for particle sizing and not the amount that the curves are shifted above the water level. The amount of shift above the water curve level is a measure of the concentration of the dispersion. It is also clear that in order obtain high-resolution particle sizing results, it is important to measure acoustic attenuation over the largest possible frequency range. From Figure 4, it is clear that good resolution between a 100 and 300 nanometer particle system could not be obtained from measurements over any narrow frequency range. As the particle size increases, the curves become more concave downwards. In the limit of very small particles, the downward concavity will only become measurable at higher frequencies. The particle size resolution with the acoustic attenuation is comparable to or better than for photon correlation spectroscopy (PCS). Neither

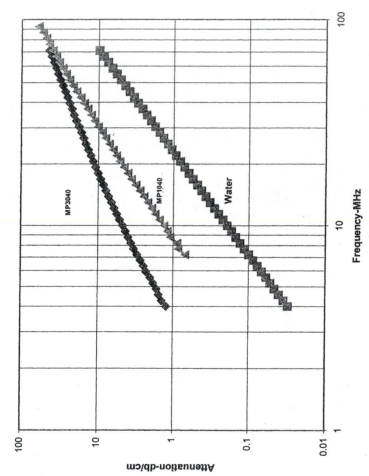

Figure 4. The acoustic attenuation spectrum of water and two different dispersed systems. Both systems are silica, 300 nanometer (MP3040) and 100 nanometer (MP1040).

PCS nor light diffraction measurements are particularly well suited in silica systems below about 100 nanometers since silica and water are very closely index matched. It is fair to say that all ensemble measurements (acoustic and light) have relatively poor resolution compared to separative methods, such as CHDF and disc centrifugation.

These measured spectra need to be converted into particle size distribution data. In light scattering, the fundamental theory used is the Mie theory (5), which relates the measured light intensities at a variety of scattering angles to a particle size distribution. The parameters that have to be put in to the Mie calculation are the real and imaginary components of the index of refraction of the particle and the liquid. In acoustics, an analogous theory has been developed, primarily by Epstein, Carhart (6), and Allegra, Hawley (7), now referred to as the ECAH theory. The ECAH theory considers the scattering of a plane acoustic wave from a collection of noninteracting particles immersed in a fluid medium. The detailed scattering computations are quite complex. The result can be viewed as a three - dimensional surface in attenuation (db/cm.) – acoustic frequency (MHz) – particle size (microns) space. Associated with each particle / liquid system is a unique surface, characteristic of that system, determined by the thermomechanical properties (density, specific heat, thermal conductivity, thermal expansion coefficient, sound speed and shear modulus {viscosity for the liquid}) of each phase. Acoustic scattering occurs as long as there is any contrast between any of the properties of the two phases. Thus for example, for a latex system where the density contrast may be zero, or very close to zero, there generally are contrasts in several other physical properties and thus acoustic scattering will still occur. ECAH theory involves computations employing often slowly converging power series expansions of complex functions. The complete solution to the scattering problem cannot be obtained in closed form so that in the published literature there are many approximations made in various limits. In our work, we employ a completely general solution developed by C.A. White, S. S. Patel, M, Carasso and J. L Valdes of Bell Laboratories, Lucent Technologies (8). This is a very rapid, robust inversion algorithm, which converts the measured attenuation vs. frequency dependence into a particle size distribution. The application of this algorithm to the MP 1040 (100 nm silica) acoustic attenuation spectra, shown in Figure 4, results in the particle size distribution (psd) shown in Figure 5. Similarly accurate results can be obtained from the MP 3040 (300 nm silica) dispersion.

Comparisons with other methods

It is instructive to compare results obtained by acoustic scattering, using the APS – 100, to results on the same sample using a completely different technique. In Figure 6 are shown the results obtained on a fume silica sample

238

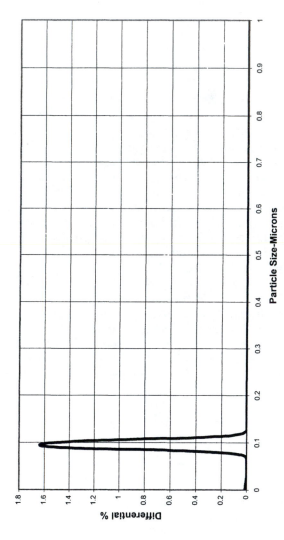

Figure 5. The particle size distribution of MP-1040 computed from the results in figure 4

using the Capillary Hydrodynamic Fraction (CHDF) technique. In Figure 7 are shown the result obtained on the same sample using the acoustic attenuation method. The results are very similar except for the broad but small peak observed at around 800 nanometers in the more concentrate sample used in the acoustic measurements. This demonstrates the utility of making particle size measurements at native concentrations. At the native concentration there is a small population of aggregates measurable whereas in more dilute systems these aggregates appear not to exist.

CMP Alumina

The particle size distribution of a fine Alumina sample is shown in Figure 8. Two modes are found at 65 and 200 nm. This Alumina product is used in CMP (Chemical Mechanical Planarization). The acoustic analysis is able to detect a small quantity of particles that are well above the size of the main (desired) particle size distribution. These larger particles are undesirable as they may cause defects due to scratching.

Particle Size Dependence on pH

The above alumina sample is manufactured and sold to CMP users at pH ~ 4. At this pH, the sample is very stable against aggregation due to the high repulsive electrical barriers between the particles. Stated differently, the Zeta Potential is large and positive at these pH values, hence the dispersion is stable. As the pH is raised by adding a base, such as KOH, the electrical barrier is lowered from a large positive to zero at around pH ~ 9.0 (Isoelectric point = zero of Zeta Potential) and then to a large negative value at pH ~ 12. At high pH values, the system is again stable against aggregation.

Figure 9 shows the particle size as a function of pH measured with the APS – 100 instrument. It is evident that at the isoelectric point (pH ~ 9) the mean particle size reaches a maximum value as would be expected. As the pH is increased to 11.5, the system disaggregates to some extent but never returns to its pH = 4 size distribution. Clearly, these results will depend strongly on the original concentration of the system and the length of time the system spends in the neighborhood of pH ~ 9. Thus, one of the utilities of acoustic measurements in concentrated dispersions is to better quantify aggregation, flocculation and gelation phenomena.

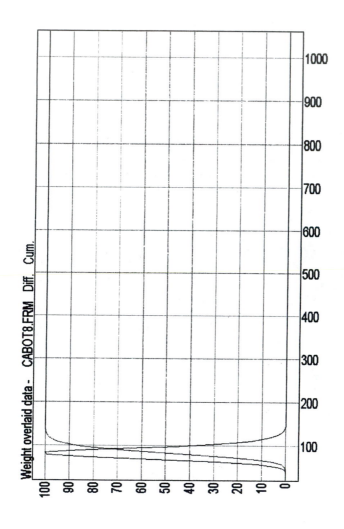

Fig. 6. The Matec CHDF 2000 results for a fumed silica sample.

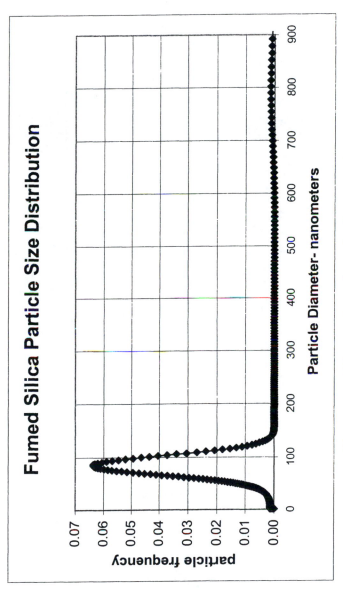

Figure 7. APS data for a fumed silica sample.

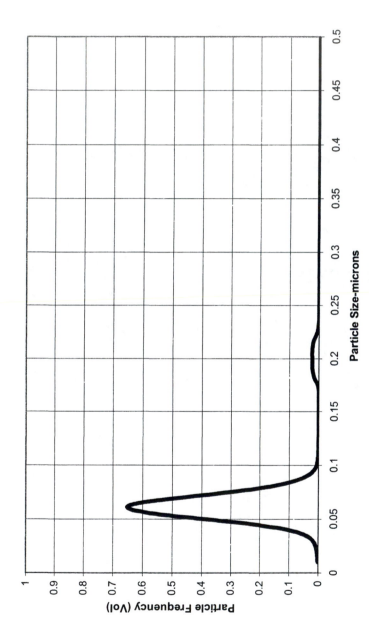

Figure 8. The particle size distribution in a CMP alumina sample

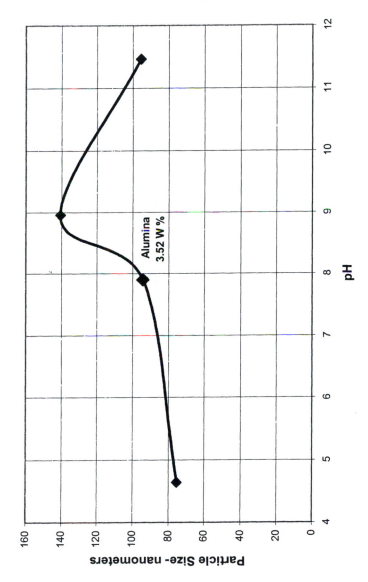

Figure 9. The mean particle size of alumina dispersion as a function of pH

Measurements of Sound Speed

Very high accuracy sound speed measurements are performed simultaneously with acoustic attenuation measurements by the APS. Therefore, it is interesting to consider the sound speed in the alumina system discussed above. In this case, we plot the change in sound speed from the sound speed at pH =4.5 and obtain a curve, Figure 10, which mimics the particle size results of Figure 9. In these measurements, it is important to control temperature to about 0.1 degree Celsius, as sound speed is strongly dependent on temperature. It is evident that one can monitor aggregation phenomena in concentrated systems by a careful measurement of sound speed.

Measurements in Emulsions

Acoustic attenuation measurements, hence droplet size distribution determinations, can be made equally well in water in oil (w/o) or oil in water (o/w) emulsions. In general it is very difficult and dangerous to dilute emulsions. Dangerous in the sense that dilution will change the equilibrium of the dispersing agent on the droplet surface to that in the liquid which in turn may alter the droplet size or may even cause creaming of the emulsion. Acoustic attenuation measurements are ideally suited for emulsion systems. Much work has been published on acoustic sizing of emulsions, generally for o/w systems, and specifically for a variety of food oils dispersed in water (9).

In Figure 11 are shown the results of a water droplet size determination in Diesel fuel, that is, a w/o system. At present, these size measurements can only be made by microscopic techniques, which are laborious and may also not be very representative of the actual system since only a very small portion of the system is sampled. The acoustic attenuation measurement is a bulk measurement, sampling a much larger volume. In Figure 11, the water droplet mean size was determined to be about 1.2 microns. The very narrow distributions below 200 nanometers may well be related to inverse micelles formed from excess dispersing agent.

Blend of Ludox TM and MP1040 Silica

In order to demonstrate the ability of the APS to identify multiple particle size populations below 100-nanometer particle size, two silica blends of Ludox TM silica (particle size ~ 30 nanometers) and MP1040 (100 nanometers) samples were prepared. Moreover, these blends were prepared using small numbers of 100-nm MP1040 particles in order to show that the APS is capable of identifying small numbers of wafer-damaging larger particles.

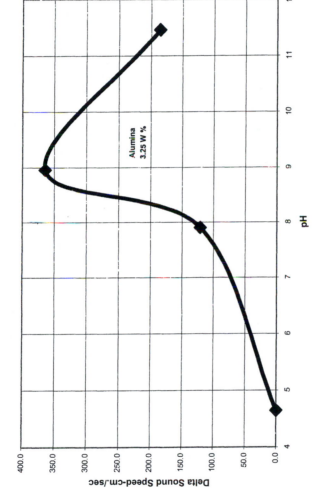

Figure 10. Measurement of sound speed for alumina system passing through the isoelectric point.

246

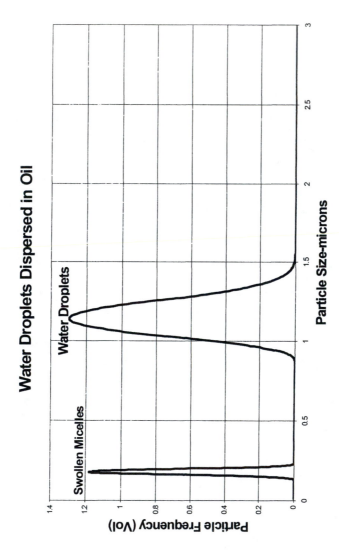

Figure 11. Acoustic determination of water droplet size in an oil phase

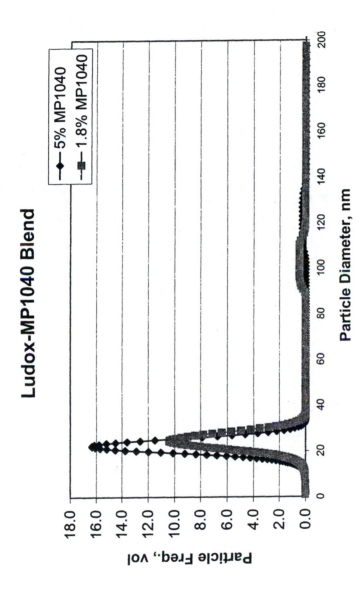

Fig. 12. Particle size distribution for two blends of 5% and 1.8% wt. MP1040 and Ludox TM particles.

Figure 12 presents particle size distribution data for these two samples. Sample "5%" was prepared using 95% of Ludox TM and 5% of MP1040. Sample 1.8% contains 1.8% on a weight basis of MP1040 particles.

The particle size distribution data clearly shows the presence of the two populations. The 100-nanometer peak is slightly shifted for the 1.8% MP1040 sample. However, this discrepancy can be considered small given the small ratio of the MP1040 particles.

Conclusions

Acoustic attenuation and sound speed measurements in concentrated systems have only recently been carried out in a routine manner. This is because high-speed computing has only become available at the desktop level and at reasonable cost in the last few years. The generation of the scattering matrix from the ECAH theory and the calculation of the particle size distributions from the acoustic attenuation data using the scattering matrix, in reasonable times, is made possible by modern computing technology. We are now in position to study the particle size distributions of concentrated colloid systems as a function of time, pH, and chemical additives. A few of these possibilities have been indicated in this article. Much new work and interesting results are anticipated in the near future.

REFERENCES

(1) Povey, M. J., *Ultrasonic Techniques for Fluid Characterization*, Academic Press, 1997.
(2) Reed, R.W., DosRamos, J.G., and Oja, T., *Review Quant. Nondestr. Eval*, **21**, Thompson, D.O. and Chimenti, D.E., Ed., pp. 1494-1501.
(3) www.matec.com
(4) U.S. Patent. Granted to Matec Applied Sciences.
(5) Bohren, C., and Huffman, D. R., *Absorption and Scattering of Light by Small Particles*, Wiley Interscience Publication, 1983.
(6) Epstein, P.S., and Carhart, R.R.J., *Acoust. Soc. Am.*, **25**, 553 (1953).
(7) Allegra, J.R. and Hawley, S.A., *J. Acoust. Soc. Am.*, **51**, 1545 (1972).
(8) Carasso, M., Patel S., Valdes, J.L., and White, C.A., US Patent 6,119,510 (19 September 2000).
(9) McClements, J. D. *Food Emulsions – Principles, Practice and Techniques*, CRC Press 1999.

Chapter 16

Stability and Structure Characterization of Cosmetic Formulations Using Acoustic Attenuation Spectroscopy

D. Fairhurst[1], A. S. Dukhin[2], and K. Klein[3]

[1]NPR Healthcare Inc., 3894 Courtney Street, Suite 180,
Bethlehem, PA 18017
[2]Dispersion Technology Inc., 364 Adams Street, Mt. Kisco, NY 10549
[3]Cosmetech Inc., 39 Plymouth Street, Suite 4, Fairfield, NJ 07004

Consistency in the stability/structure characteristics of an emulsion is a necessary prerequisite to achieve the desired final product quality. Cosmetic emulsions are complex, multicomponent systems that, invariably, cannot be diluted without consequence. Current non-imaging instrumentation based on light scattering methods is severely limited in its application to such systems. We report the results of a preliminary study of a variety of "real-world" cosmetic emulsions, both O/W and W/O, from sunscreens to moisturizers measured without dilution using a technique based on acoustic spectroscopy in combination with electro-acoustics. A major advantage of the technique is the ability to study emulsions under flow conditions, allowing measurements under different shear conditions. In addition, a unique feature is the ability to non-destructively probe structural characteristics of emulsions. This important development will be discussed in light of its correlation with classical rheological (viscosity and elasticity) measurements.

The design of a cosmetic emulsion is quite complex and often requires several iterations to produce a stable, efficacious, safe, cost effective and elegant product. While product specifications are established to ensure product consistency during manufacture, all too often they are not sufficient to guarantee that a truly high quality product has been manufactured. What is needed is an objective test that will be easy to run, and predictive - a test that measures *fundamental* characteristics (1,2). Any test must, however, not be a function of the instrument used nor of the operator.

Two reliable and well-established parameters are, respectively, particle (droplet) size and ζ potential (surface charge) (3,4). The techniques devised to determine them are extremely diverse. Each method has its attractions (5,6) but it is important to select the one that meets the actual requirements of the application in question; versatile as many current instruments may be, the one that does everything has yet to be invented. Unfortunately, the vast majority of current, commercially available, instrumentation is based on light scattering. This severely limits its application to cosmetic emulsions because of the requirement to dilute the system under investigation.

Cosmetic emulsions are multi-component systems that, invariably, cannot be diluted without consequence. The (dilution) process not only destroys any structure characteristics but also can induce instability leading to creaming and, often, breaking/coalescence. Hence the oft-found difficulty of correlating particle size (PS) and zeta potential (ZP) measurements with flow properties, freeze-thaw and shelf storage behavior. Additionally, many instruments are only suitable for studying O/W emulsions, or may have difficulty evaluating emulsions that contain significant concentrations of dispersed particulates, as may be found in sunscreen formulations containing zinc oxide/titanium dioxide.

It has been long known that ultrasound based techniques open an opportunity to eliminate dilution (7-10). However, instrumentation hardware was too complex and suffered from weaknesses in practical application of the technique to systems of high internal phase concentration. In addition, the theoretical basis of the deconvolution of the raw data was not capable of dealing with particle-particle interaction - an essential feature of concentrated dispersions. Recently, all these problems have been addressed and a commercial instrument developed that convincingly demonstrates the advantages of a technique based on a combination of acoustic attenuation and electro-acoustic spectroscopy (11,12). This offers a new opportunity for characterizing these complicated cosmetic systems; it has the advantage of being able to study turbid emulsions where other techniques, such as light scattering, simply will not work and where dilution of such systems can alter their physical properties. It is also sensitive enough for characterizing polydispersity. A major advantage of any technique based on sound is that a sample under investigation need not be stationary - a necessary requirement of light scattering instrumentation. Hence, an acoustic instrument holds promise of being able to study emulsions that are flowing, in-line or on-line. The technique has been successfully applied to the study of simple (3 component) model emulsions (13,14). Unfortunately, model systems, while of academic interest to the researcher, are often of little practical use to the cosmetic formulator.

In this paper we report the results of a preliminary study of a variety of "real-world" cosmetic emulsions, both O/W and W/O, from sunscreens to moisturizers. The aim of the study was to determine the limitations of the acoustic method and to show proof-of-concept in the ability of the technique to help characterize the stability and performance of cosmetic emulsions without the need for dilution.

Experimental

A series of emulsions were prepared; they were formulated to be commercially viable cosmetic products. The details of the systems are given in Table 1. The emulsions varied widely in composition and character viz. internal phase weight fraction, viscosity and stability.

Table 1. Cosmetic Emulsion Formulations
Systems formulated to be commercially viable products

Product	Emulsion Type	Emulsion Stabilizer	Internal Phase Wt. Fraction	Viscosity (cP)	Number of Components
A	O/W	Anionic	19%	13,800	9
B	O/W	Non-ionic	16%	4,600	14
C	O/W	Non-ionic	27%	6,400	18
D	O/W	Non-ionic	32%	4,850	13
E	W/O	Non-ionic	74%	134,000	10

The instrument used was a DT-1200 Acoustic Spectrometer developed by Dispersion Technology, Inc., Mt Kisco, New York (15). This instrument has two separate sensors for measuring acoustic and electro-acoustic signals independently; details of the instrument and its operation are given elsewhere (13,14,). The raw data output is ultrasound attenuation frequency spectra within a frequency range, from 1MHz to 100MHz, that is required for characterizing colloid systems with typical size heterogeneity from nanometers to hundreds of microns. The attenuation spectra contain information about particle size distribution (PSD) and high frequency longitudinal rheology of the sample. Because the power of the applied ultrasound pulse is very low, the technique is truly non-distructive; it does not affect the stability of a suspension. It is suitable for characterizing hard solid particles as well as soft emulsion droplets and laticies.

The electroacoustic sensor raw data output is Colloid Vibration Current; this measured parameter yields information about ZP. Complete PSD and ZP measurements can be completed in minutes and can be made on flowing systems.

Results and Discussion

Figure 1 gives the results of the raw attenuation data for Emulsion A, an anionic (glyceryl stearate + AMP stearate), basic O/W cosmetic emulsion. The curve A corresponds to an initial measurement of the emulsion at rest whilst the curve B is the result after subjecting the emulsion to shear produced by peristaltic pumping for 30 hours. Clearly, there is a marked change in attenuation indicating that the particle size distribution (PSD) has changed; the emulsion is not "stable" to the level of shear produced by the peristalsis. Figure 2 shows measurements made, at two different points on the attenuation curve (at frequencies 4MHz and 100MHz), as a function of time of shear. These time dependences are different and this reflects a variation of different droplet sizes.

The low frequency attenuation (4MHz) drops initially rapidly and then levels off (curve B). Attenuation at this frequency is mostly dependent only on the large droplets. The observed time dependence of an initial rapid drop in attenuation indicates that the shearing action has destroyed large size droplets.

The reduction in high frequency attenuation (100MHz) is much slower (curve A). Attenuation now depends on both small and large droplets. The longer time dependence of the high frequency attenuation is what would be expected if the small droplets were removed by, for example, coalescence. This is consistent with the removal of large droplets at the initial much shorter times.

Thus, we conclude that the original emulsion must be bimodal and the shearing action produces two effects: large droplets are broken down into smaller droplets and smaller droplets are coalescing. The overall result is depicted in Figure 3 showing the calculated PSD for the emulsion before and after shear. Initially, the emulsion comprises primarily very small droplets (ca 70nm mean) with a fraction of droplets about 8um. After shearing, the PSD changes; there are fewer of the small droplets and considerably more larger droplets but the mean PS of the larger fraction is now around 4um. Thus, it is clear that, using acoustic measurements, dynamic changes in the state of an emulsion can be followed in real time

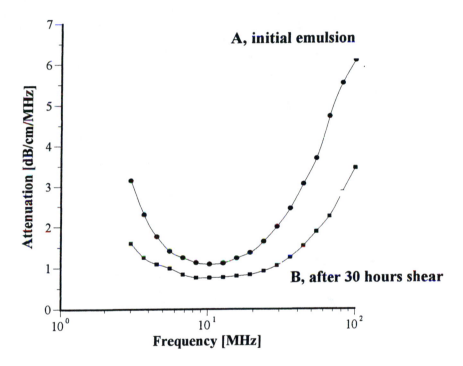

Figure 1. Effect of shear on the attenuation spectrum of an O/W anionic emulsion.

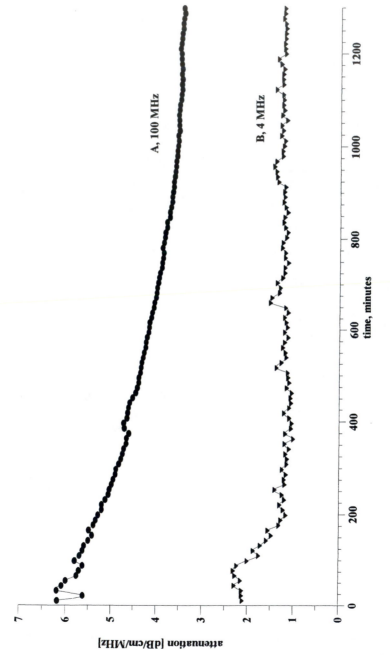

Figure 2. Change of attenuation, at 4 MHz and 100 MHz, as a function of time of shear of the anionic emulsion.

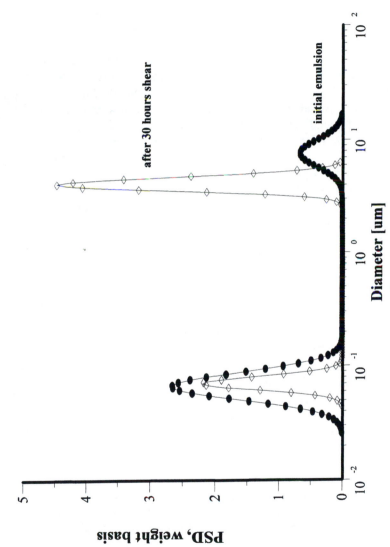

Figure 3. Effect of shear on the droplet size distribution of the O/W anionic emulsion.

The original emulsion was found to have a ZP of -20.0mV, (determined utilizing the electro-acoustic feature of the DT-1200), a not unexpected value given the known facts of the emulsion; it is quite typical of such anionically stabilized systems (5).

Emulsions B and C are examples of nonionically stabilized O/W moisturizers; primarily they differed in that Emulsion C also contains 3.5wt% of an inorganic particulate sunscreen active - a microfine hydrophobically coated zinc oxide (ZnO) added to provide broad spectrum UV protection. The raw attenuation spectra are compared in Figure 4; they are clearly distinguishable from one another and it is also apparent that the shapes of the two spectra are dissimilar. Note also, the good reproducibility of the technique. The PSD of both the oil droplets and the ZnO, dispersed therein, can be abstracted from the total spectrum for Emulsion C (Figure 5). The values of 1.92μm and 165nm are typical for emulsion oil droplets and ZnO particles respectively. Indeed, the ZnO size is very good agreement with the mean PS of 148nm, measured on a predispersion of the ZnO in the oil phase prior to emulsification. In this latter case the instrument used was an X-Ray disc centrifuge sedimentometer – a high resolution PS analyser. The technique is ideally suited to the measurement of the PS of oxides and ceramics (16) but is limited to working with concentrations less than 2vol%.

The Emulsion C was then re-formulated using 3.5wt% of a basic USP grade ZnO, such as might be used in a baby diaper cream, in place of the microfine-grade ZnO. USP grade ZnO has a much larger mean PS and a much broader PSD than the microfine grade material. The results are compared in Figure 6. The mean PS calculated for the USP ZnO was 0.55μm; again in very good agreement with that measured for a straight aqueous dispersion of the material (0.48μm). Correct predispersion of a particulate sunscreen active, such as ZnO or TiO2, into the oil phase of an emulsion is critical to obtaining optimum performance (i.e. minimum whitening with maximum SPF per % active)(17). The results of the present study suggest that acoustics might be useful in measuring the state of aggregation of ZnO or TiO2 prior to emulsification thus allowing a formulator to monitor the efficiency of the homogenization process. It should also find utility in the milling and grinding of pigments used in color cosmetics.

Figure 7 shows the attenuation spectrum for the emulsion D, a stable, O/W waterproof sunscreen formulation. The attenuation was measured first under quiescent conditions and then after the emulsion was deliberately destabilized. The instability was induced by simply diluting *in-situ* approximately 2.5 times with deionized water. It can be seen that the attenuation drops by a factor 10, reflecting the observed coalescence and phase separation. Again, the results demonstrate the ability of the acoustic method to characterize the behavior of

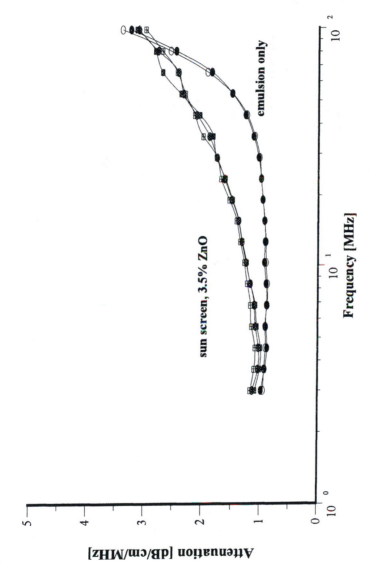

Figure 4. Comparison of O/W moisturiser formulations.

258

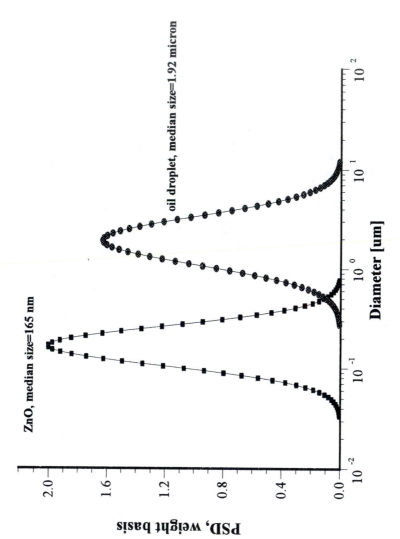

Figure 5. Particle size distribution of ZnO suspension within the Oil phase of an O/W moisturizer.

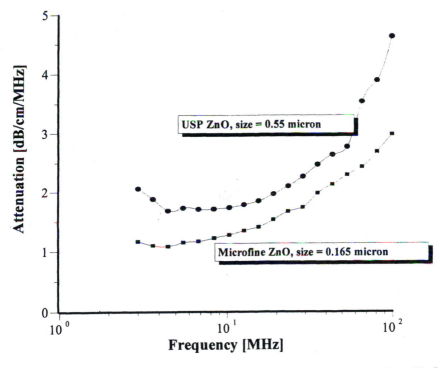

Figure 6. Comparison of moisturizer formulated with different grades of ZnO.

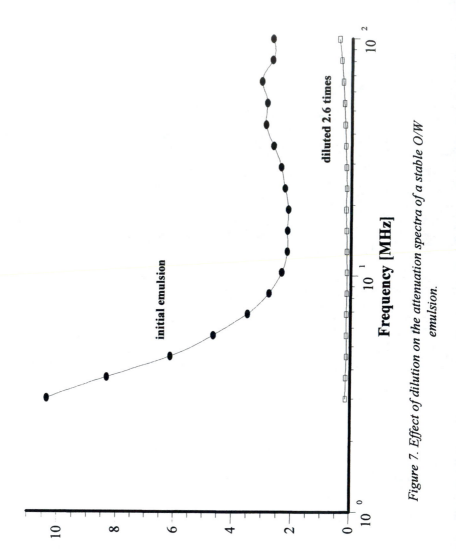

Figure 7. Effect of dilution on the attenuation spectra of a stable O/W emulsion.

an emulsion under varying chemical conditions in real time and under dynamic conditions.

It is interesting to note that, for this system, the measured ZP value was +4.5 mV. This is surprising as it had been assumed that the emulsion was nonionically stabilized because Laureth-23 had been used as the emulsifier and, typically, nonionically stabilized suspensions have a ZP value close to zero mV but negative in sign (5). On examination of the formulation components for this emulsion, it was found to contain EDTA and TEA. Thus, the only explanation for the positive value of the ZP is that the either, or both, of the amines have specifically adsorbed at the oil/water interface. This result warrants further investigation but nevertheless demonstrates the usefulness of acoustics to determine and confirm the charge nature of emulsions.

The W/O emulsion E was a classic (beach type) sunscreen formulation that was found to be unstable and, subsequently, it separated into two distinct phases. Pumping the emulsion into the DT-1220 allowed the two phases to simply mix; the resultant attenuation spectrum is shown in Figure 8. An emulsion PS of 40um was calculated from the initial attenuation measurement (curve A). Addition of 2% sorbitan laurate (curve B) did little to improve the appearance or stability; the calculated PS remained around 40um and the "emulsion" was very thin and of low viscosity. Increasing the surfactant concentration to 5.5% (curve C) resulted in a dramatic change to a white emulsion of very high viscosity having a mean PS of only 2um.

The forgoing studies strongly suggest that an acoustic measurement might be used as a formulation aid to "fingerprint" emulsion composition. This hypothesis is further illustrated by comparing the attenuation spectra of three different "components" that would typically be produced during the course of formulation or manufacture of a cosmetic. These were an "oil phase", the same oil phase containing predispersed 14wt% ZnO and the final sunscreen following addition of water and homogenization. The spectra are clearly different (Figure 9) and track the compositional nature. There is an increase in attenuation of the oil phase upon addition of the particulate ZnO because, as a solid, it has a larger attenuation than any liquid. However, the final sunscreen has the smallest attenuation; it is considerably less than the oil phase even though it contains ZnO. This is primarily because we have added a large concentration (ca 60wt%) of water and this has the smallest attenuation (almost zero) of any liquid. In addition, the oil/ZnO attenuation is also significantly reduced because the concentration of that component is now only about 35wt%; hence, the actual ZnO concentration in the sunscreen is only 3.5wt%. This suggests, therefore, that it is possible to use the attenuation spectrum as a measure of reproducibility and repeatability at any stage of formulation; this, in turn, would considerably improve the consistency of the final product and, overall, lead to a cost benefit.

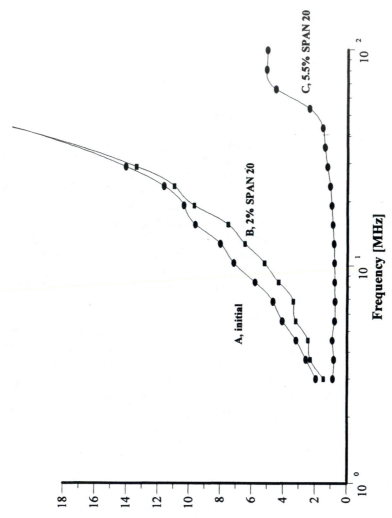

Figure 8. Attempt to re-stabilize an unstable (separated) W/O emulsion by addition of emulsifier.

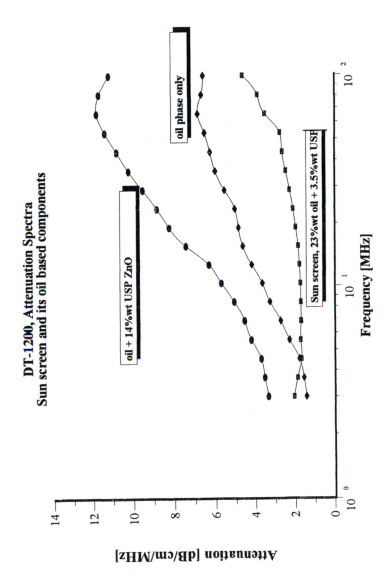

Figure 9. Fingerprinting Emulsion Composition.

During the course of background research into the use of acoustics it was found that, from the measurement of both ultrasound attenuation and speed, it is possible to obtain information about the microscopic structural characteristics of fluids and suspensions. Indeed, there are dozens of papers devoted to pure liquids and polymer solutions. Despite this, acoustics remains, practically, an unknown tool with which to characterize the rheological behavior of emulsions. A thorough review of the whole of this subject has recently been made by Dukhin (18). Importantly, the technique is non-invasive and non-destructive, hence it should be possible to probe even extremely weak structures. In addition, no extra preparation or measurements are needed; the rheological data are calculated from the same attenuation spectra used for PS analysis.

Figure 10 shows the measured attenuation of a series of pure liquids as a function of frequency. The relationship is linear as predicted by simple theory (19) for Newtonian liquids. The viscosity value consequently calculated for a few selected liquids is given in Table 2.

TABLE 2. Viscosity of Pure Liquids

Liquid	Viscosity (Nominal)	Viscosity (Calculated)
Kerosene	4.7cP	4.6cP
Silicone Fluid A	4cSt	4.3cP
Silicone Fluid B	10cSt	12.0cP

There is good agreement with the nominal (literature) values, even after converting the dynamic viscosity (St) to kinematic viscosity (P). Next, a typical oil phase of an emulsion was prepared; this contained various components including waxes and emulsifiers. The measured attenuation is shown in Figure 11. Though the attenuation is not perfectly linear, the calculated viscosity of 35cP compares favorably with the 37cP measured, at the same temperature, using a Brookfield viscometer. This, however, may be fortuitous because the "viscosity" measured using an ultrasound technique is based on a longitudinal shear whereas that measured using a conventional rotational viscometer, like the Brookfield, involves a transverse shear. Essentially, a rotational viscometer measures a low frequency loss modulus whereas an acoustic "viscometer" gives a high frequency loss modulus. Hence, the presence of any "dispersed phase" in a liquid will result in non-linearity and a disparity with a "normal" viscosity. Indeed, this will also be the situation if the (pure) liquid has any microscopic structure. This can be clearly seen in Figure 12 which shows the loss modulus, G", determined from the attenuation spectrum for the two ZnO-based

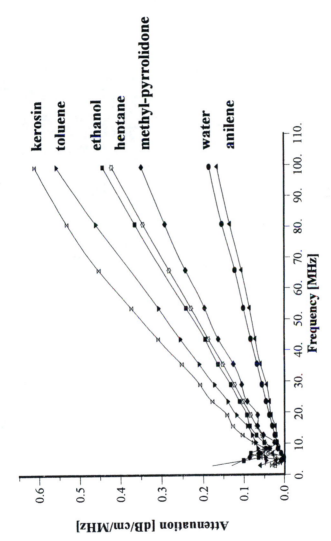

Figure 10. Attenuation Spectra for various pure liquids.

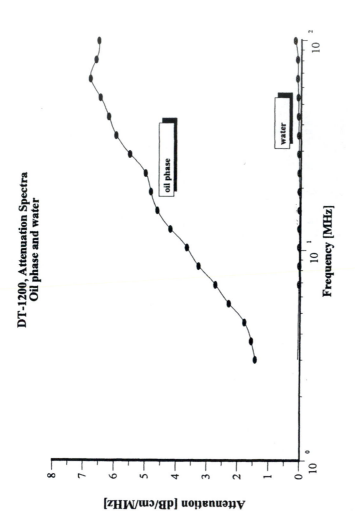

Attenuation of water at 100 MHz = 0.2   Viscosity of water at 25°C = 1 cP

Attenuation of oil phase at 100 MHz = 7 Viscosity of oil phase = 35 cP

Figure 11. Attenuation spectrum of the oil phase of an emulsion.

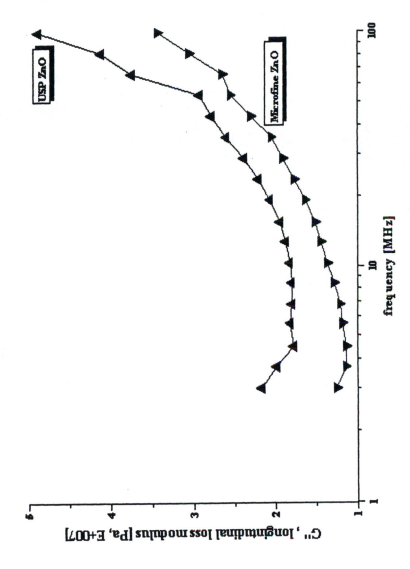

Figure 12. Loss Modulus, G", calculated from attenuation spectra, for two sunscreen formulations.

sunscreens (Figure 5). It is clear that the response is non-linear and, hence the system is non-Newtonian. However, it is still possible to calculate a "viscosity" using the measured attenuation and assuming the simple linear relation between G", frequency, ω, and viscosity, η (19). This being so, the viscosity of each sunscreen was determined to be 5600cP and 7900cP for the microfine ZnO sunscreen and USP ZnO sunscreen respectively. These values are surprisingly close to those measured using a Brookfield (5000cP and 7000cP); indeed, the viscosity ratio is 1.41 (acoustic) and 1.40 (Brookfield) respectively. The only explanation is that there must be some microscopic structure (and no oil droplet-oil droplet interaction) in either emulsion. The observed increase in viscosity is pure hindrance arising from the PSD. The presence of the ZnO particles (of any size) within the oil droplets appears to be irrelevant. This warrants further study.

Conclusions

Initial results, from studies of a series of commercially formulated cosmetic emulsions, suggest that a technique based upon a combination of acoustic attenuation and electro-acoustic spectroscopy is a powerful new tool to characterize the stability and performance of cosmetic emulsions without the need for dilution. Measurements are made in real time and can follow dynamic changes as the system is perturbed. Particle (droplet) size and zeta potential can be easily measured on both O/W and W/O systems; as long as the emulsion can be pumped it can be measured. Indeed, for the systems in the present study, it should be possible to measure them in-line – a potential major advantage for manufacturing and production. It appears feasible to use the attenuation spectrum as a measure of reproducibility and repeatability at any stage of formulation; this, in turn, would considerably improve the consistency of the final product and, overall, lead to a cost benefit and savings.

It is difficult at present, given the preliminary nature of the study, to assess the precise usefulness of rheological data from acoustic measurements. However, it is clear that the technique offers an additional route to probe the microscopic structure(s) in emulsions and suspensions. Importantly, the technique is non-invasive and non-destructive, hence it should be possible to probe extremely weak structures. In addition, no extra preparation or measurements are needed; the rheological data are calculated from the same attenuation spectra used for PS analysis.

Based on the present work many avenues of investigation suggest themselves. Future studies will probe in more detail the ability of the technique to discriminate the presence of a particulate phase within the emulsion and then to follow this as a function of homogenization. It would be interesting to determine, for a given emulsion, the effect(s) of HLB of emulsifier and the addition/variation of thickeners, viscosity/rheology modifiers.

Literature Cited

1. Lyklema, J. *Fundamentals of Interface and Colloid Science*, Volume 1, Academic Press: 1993.
2. Fairhurst, D. *Tire Technology International*, 151-154 (1996).
3. Stanley-Wood, N.G.; Lines, R.W., Eds.; *Particle Size Analysis*, Royal Society of Chemistry: 1992.
4. Hunter, R.J. *Zeta Potential in Colloid Science*, Academic Press: 1981.
5. Barth, H.G., Ed.; *Modern Methods of Particle Size Analysis*, Wiley Interscience: 1984.
6. Provder, Th. Ed.; ACS Symposium Series **332** (1987), Series **472** (1991) and Series **693** (1996).
7. Marlow, B.J.; Fairhurst, D.; Pendse, H.P. *Langmuir*, **1983**, *4*, 611-626.
8. McClements, D.J. *Colloids and Surfaces*, **1991**, *37*, 33-72.
9. Alba, F.; Dobbs, C.L.; Spark, R.G. *First International Particle Technology Forum*, Part 1, 36-46 (1994).
10. O'Brien, R.W.; Cannon, D.W.; Rowlands, W.N. *J.Colloid and Int. Sci.*, **1995**, *173*, 406-418.
11. Dukhin, A.S.; Goetz, P.J. *Langmuir*, **1996**, *12*, 4336-4344.
12. Dukhin, A.S.; Goetz, P.J. *Colloids and Surfaces*, **1998**, *144*, 49-58.
13. Wines, T.H.; Dukhin, A.S.; Somasundaran, P. *J. Colloid and Int. Sci.*, **1999**, *216*, 303-308.
14. Dukhin, A.S. *et al*, Colloids and Surfaces A, **2000**, *173*, 127-159.
15. Dukhin, A.S.; Goetz, P.J. U.S. Patent 6109098, 2000.
16. Fairhurst, D.; McFadyen, P. *Clay Minerals*, **1993**, *28*, 531-537.
17. Fairhurst, D.; Mitchnick, M. In *Sunscreens: Development, Evaluation and Regulatory Aspects,* Lowe, N.J.; Shaath, N.A.; Pathak, M.A., Eds.; Marcel Dekker: 1997, Second Edition.
18. Dukhin, A.S.; Goetz, P.J. In *Ultrasound for Characterizing Colloids. Particle sizing, Zeta Potential, Rheology*, Elsevier: 2002.
19. Temkin, S. *Elements of Acoustics*, John Wiley: 1981, First Edition.

Indexes

Author Index

272

Subject Index

284